Motorbooks International

POWERPRO SERIES

HOW TO BUILD & MODIFY PISTONS, RODS & CRANKSHAFTS

Ben Watson

To: Anne Marie, Gladys, Bennie, William, Sam, Pamela, Hartman, Beverly, K.T., Virginia, Burt, Andrea, Gilda and Benjamin Franklin McLinton.

First published in 1996 by Motorbooks International Publishers & Wholesalers, 729 Prospect Avenue, PO Box 1, Osceola, WI 54020-0001 USA

Library of Congress Cataloging-in-Publication Data

Watson, Ben.
How to build & modify pistons, rods & crankshafts / Ben Watson.
p. cm. -- (Motorbooks International powerpro series)
Includes index.
ISBN 0-7603-0079-8 (pbk.)
1. Automobiles--Motors--Pistons and piston rings.
2. Automobiles--Motors--Crankshafts.
I. Title II. Series.
TL214.P57W38 1996
629.25'2--dc20 96-14056

On the front cover: A few high-performance parts for a Chevrolet, 0.030in-over, 350ci engine. The crankshaft and 6in connecting rods are manufactured from 4340 steel, the pistons are forged aluminum, and the bearings are lead indium. The piston rings consist of a plasma moly top ring, a Zero-Gap second ring, and a standard three-piece tension oil control ring. All parts courtesy of Childs & Albert, Inc. of Valencia, California. *David Gooley*

Printed in the United States of America

Contents

Introduction

Prologue

When I was more concerned with fast cars and cruising the 1849 Village than with Senior English at Arlington Heights High School, I thought the ultimate powerhouse would consist of "pop-up pistons," a stroker/roller crank, and a high-compression ratio. As the years passed, I became wiser and controlled vast wealth supplied by a $2.25 an hour job. A close high school friend and I began to build a race car. We bought a $250 Mustang from a used car lot on the north side of town. After cutting off the old brake drums, installing new brakes, and painting the car by hand with a paintbrush, we were ready to race. We knew from the first trip to the track that we were going to need a true racing engine if we were going to be successful.

A fellow I worked with at the time mentioned that he had a 302 Ford engine which he was going to build into a racing engine. Fortunately for us, a recent marriage and the obligatory associated financial problems required him to curtail the endeavor. Picking the engine up for a mere fraction of what our youthful vanity felt it was worth, we began to build our racing engine. The pistons supplied with the engine had an impressive "dome" on them. The block had been somehow damaged, warped, and distorted. When the main bearing caps were torqued to the proper specs, the crankshaft would no longer rotate. The main bearing journals were misaligned. When the engine was line bored, the position of the crankshaft was closer to the top of the engine block deck. In fact, it allowed the piston to protrude 0.050 inches (in) above the top of the block. This meant that even with a fiber head gasket installed, the piston would arrive at top dead center and squarely smack the bottom of the cylinder head. When we used a thin metal head gasket, the piston solidly hit the bottom of the head. In spite of the "negative" piston-to-head clearance, we had accomplished our goal of maximizing the compression ratio. In fact, years later when I had learned how to calculate such, I learned that the compression ratio was well over 16:1.

For some, performance is the measurement of the amount of chrome, glitz, and polish the vehicle or engine has. This book looks at the serious side of performance, cuts through the bench racer's myths, and helps the reader build maximum performance at minimum expense.

While our goal of maximized compression ratio was achieved, the emphasis on this single aspect of engine building showed the limits of our knowledge at the time. An engine is like a pump, with a couple of twists. It must draw in air and fuel during the intake cycle, compress the mixture, ignite the mixture, and expel the waste gases. Power from an engine depends on its ability to perform each of these functions well.

This book is intended to help the reader prevent such tragedies in their own lives.

The history of the spark-ignition engine

The internal-combustion engine, like any other technological innovation, did not just happen. The engine that pushes your car down the road was the end result of centuries of technological progression. When we think about the internal-combustion, spark-ignition engine, commonly referred to as the gasoline engine today, it is difficult to think about the theoretical research and development that brought it about. Today, the gasoline engine permeates every aspect of our lives. We travel to work or to the grocery store with it, we cut our grass with it, we fly aircraft with it, and we eat weeds with it. However, in the mid-nineteenth century the spark-ignition engine was little more than a really good idea.

The story begins in ancient China 221 years before the birth

The name of this store south of Seattle implies something about the automotive performance parts industry. I can remember as a neophyte engine builder in the late 1960s and early 1970s sitting with a performance catalog and thinking about the engine I was about to build. The stock engine was 225hp.This clear distributor cap will add 5 percent to the horsepower. That set of headers will add 15 percent. This set of pistons will add 25 percent. That camshaft will add 15 percent. And this set of spark plugs will add 5 percent. The engine horsepower will, therefore, be 411. This "performance catalog estimate" completely ignores the laws of physics. The truth, however, is that if any of these items were added as part of a coordinated effort to improve the performance of the engine, then the advertised performance improvement might be achieved.

Once an American male rite of passage, today's emissions laws mean that the young men of the 1990s are stuck doing performance modifications to engines and cars that their fathers may have considered modifying. Future generations will be forced to find rites of passage in less meaningful endeavors. This early 1990s Buick Regal featured outstanding performance "right out of the box," but if the performance industry of the 1990s were on a par with that of the 1960s, how long would those front tires last?

This is the hood of a 1917 McLaughlin, a Canadian Buick. When this car was built, the internal-combustion, spark-ignition engine had already been in serious development for over fifty years. Fifty years later, 1967, the art of performance engine building, for the amateur, reached its zenith.

Even as early as 1917, the basic characteristics of the modern performance engine were in place. Overhead valves, secondary distribution ignition, the search for the most efficient crankshaft configuration. Even in that day the in-line six-cylinder engine was hailed for its efficiency of design.

of Christ. Ancient records indicate that during the Chin dynasty men curious about the nature of substances found in and on the earth discovered that when sulfur, potassium chloride, and charcoal were combined in a finely ground powder, it formed a substance that, when ignited, created a hot, rapidly expanding gas. We refer to this substance as gunpowder. With this single development the paths to an incredible array of technologies began. One of these paths would lead us eventually to the hydrogen bomb. We will begin our journey of innovation on that path but will leave it at a fork along the way.

For over a thousand years gunpowder did little more than impress the guests of the emperor of China, a simple party trick. Then someone in the Chinese imperial court got the great idea of placing the gunpowder in a hollow tube with one end plugged. When the gunpowder was lit, it propelled the tube high into the air. It took a European to figure out how to kill with it in a really impressive way. In 1242, an Englishman from Oxford University named Roger Bacon published the gunpowder formula. As with most university-trained eggheads, he did not see what a really neat tool this would be for destroying large amounts of property and people. This event had a historical significance on par with the accomplishment of the Rosenbergs in the late 1940s.

In the thirteenth century, one of the problems of making war was that if you took the war to the enemy he would often protect himself within the stone walls of a castle. Many clever generals and their engineers designed many devices to hurl large stones at the walls. The

Time line in the development of the spark ignition engine

- B.C. China - "Gunpowder" invented
- 1242 - Europe discovers how to kill efficiently with it
- 1260 - Wooden cannon developed
- 1326 - Metal cannons developed
- 1585 - "Timed bombs" developed
- 1654 - The nature of gases first explored scientifically
- 1758 - The multiple barrel musket
- 1790 - The ball bearing
- 1853 - The first "gas" engine
- 1885 - Gas-engine automobile
- 1994 - Jacque Villeneuve wins Indianapolis 500

The spark-ignition engine that we take for granted today has a long history of development. Not all of the development process occurred in a straight line. One individual would find the solution to a problem; a second individual would see that solution as the basis for resolving a problem that the first individual never dreamed of, and so on.

HALF PRICE BOOKS®

Half Price Books
1835 Forms Drive
Carrollton, TX 75006
OFS OrderID 39900017

Thank you for your order, ABE APTP Order!

Thank you for shopping with Half Price Books! Please contact Support@hpb.com. if you have any questions, comments or concerns about your order (741754978)

Visit our stores to sell your books, music, movies games for cash.

SKU	ISBN/UPC	Title & Author/Artist	Shelf ID	Qty	OrderSKU
S477886865	9780760300794	How to Build & Modify Pistons, Rods, & Cra... Watson, Ben	TRN7.2	1	

SHIPPED STANDARD TO:
ABE APTP Order
800 Avondale Ave
JR4FZVDY
Columbus OH 43212
help@textbookrush.com

ORDER# **741754978**
Abebooks

idea was to bombard the walls with heavy stones until they fell. Catapults and battering rams worked well up close and with a lot of repetition. Something was needed to increase the damage done by the stones, a more forceful way of propelling them for instance. One story holds that in 1260 an otherwise unknown figure in history, Konstantin Anklitzen, decided that if the rapidly expanding gases of burning gunpowder can propel a cylinder, then it should also be possible to mount the cylinder solidly to the ground and used the gunpowder to propel a projectile. His "cannon" is said to have been made of wooden staves banded by iron. It resembled a wine barrel more than it did a cannon. Humankind might have been better served if the device had been used as a wine barrel. A little over sixty years later, in 1326, metal cannons were used. At this point, three principle components of the internal-combustion engine were in place: the rapidly expanding gas (gunpowder), the cylinder (cannon), and the piston (stone cannonball).

In 1585, the Dutch army was engaged in a siege of Antwerp. By this time fortifications designs had been revised to be resistant to the cannon. A way had to be devised to precisely position an explosive device. This, however, could be embarrassing and fatal to the poor soul placing and igniting the explosive device. The solution was to design a bomb so that it would detonate at a later time. We now have the principle of remotely and mechanically kindling an explosion.

A few years later, in 1654 Germany, Otto von Guerick designed a pump that could be used to remove air from a glass jar. The result of this development was that others could explore gases, air pressure, and the movement of air between areas of varying pressure and the behavior of gases as pressure changes. This led to a better understanding of how a projectile is propelled with expanding gases and how air might be drawn into an area recently vacated by a descending piston.

In 1718, James Puckle proposed a multiple-barrel musket that would fire each barrel sequentially. Later refinements of this concept during the American Civil War led Richard Jordan Gattling to invent the weapon that bears his name. Of course we are not talking about weapons here, but the concept of automatically sequencing small explosions is fundamental to a multiple-cylinder engine.

Later in the eighteenth century, the ball bearing was developed by Philip Vaughan. Thirteen years later, in 1807, Isaac de Rivez received a patent for a gas-driven automobile. In 1853, Eugenio Barsanti and Felice Matteucci announced the development of the gas engine. Gottlieb Daimler invented the surface-feed carburetor in 1876; this evolved into the float-feed carburetor of Edward Butler in 1884, which led to the gas-engine automobile of Gottlieb Daimler, Wilhelm Maybach, and Karl Friedrich Benz in 1885. From here it was a short jump to the 1994 Indianapolis 500 victory of Jacque Villeneuve.

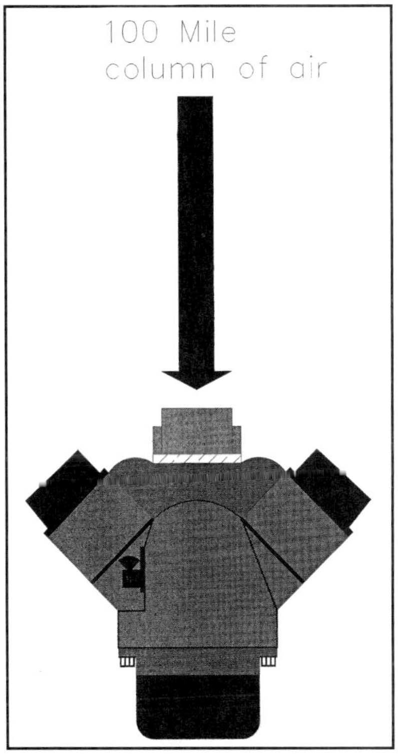

There is a 100-mile column of air that extends from the surface of the Earth to the domain of the satellites. The goals of the performance engine builder is to persuade a small portion of this column into the combustion chamber of the engine he or she is building. Many people, the ones who have never really thought about it, think that engine performance and power comes from fuel. Getting the fuel into the engine is the easy part. Getting the air into the engine, regardless of the speed of the engine, regardless of the fuel metering system, regardless of the air filtration and induction system . . . that is the hard part.

A little bit about air induction

The hundred-mile column

Above the earth is a column of air that rises more than 100 miles. Those who know about the world, the purpose of the universe, and the meaning of life know that this column of air was placed here by higher powers for one reason only: to power automobiles. If one were to approach Earth from space, one would find that as he or she descended through the atmosphere it would exert an ever-increasing force on the spacecraft. By the time our space traveler were to land on the surface near Galveston, Texas, he or she would find that force to be nearly 15 pounds per square inch (psi) of pressure. The sole purpose of this pressure is to force air into the intake system of fire-breathing internal-combustion engines.

Contrary to popular belief, the internal-combustion gasoline engine does not draw air into its bowels when it runs. Instead, the piston descending in the cylinder stretches the molecules in that cylinder apart. As the molecules

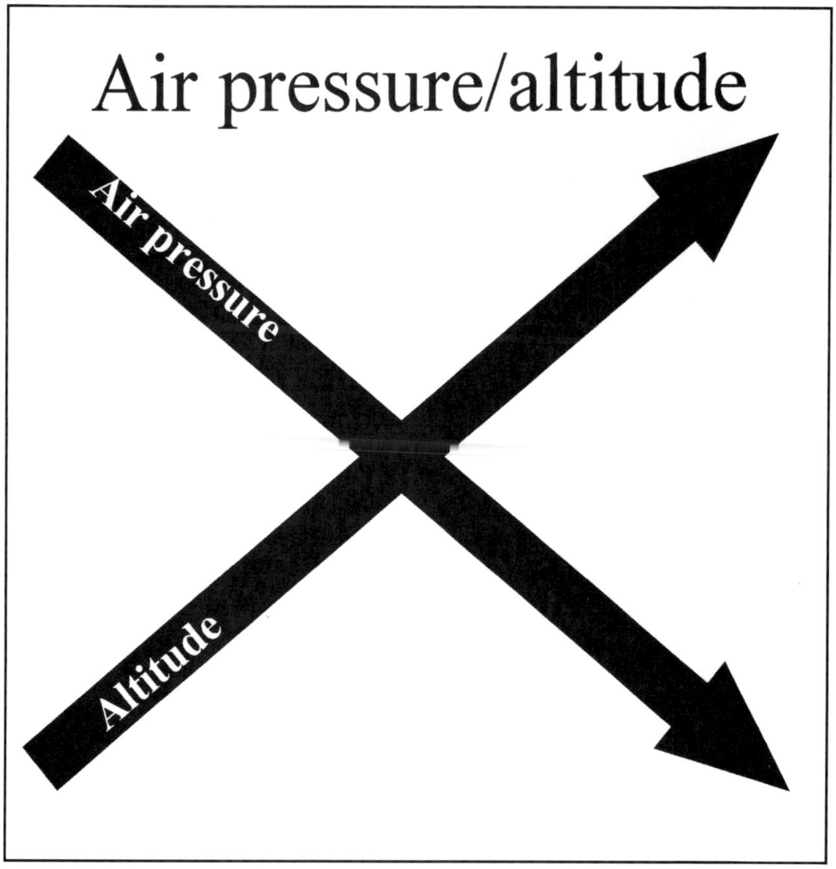

There are many external things that can affect the flow of air into the intake and combustion chamber; therefore, there are many external things that can affect the performance of the engine. As the altitude the engine is being operated at increases, the amount of air molecules in the combustion chamber decreases. The result is lower potential power at high altitudes. This can be compensated for with supercharging or turbocharging.

are stretched apart, their pressure decreases. When the pressure in the cylinder drops below the pressure exerted by the 100-mile column of air, the air in that column begins to flow into the cylinder. The more the pressure in the cylinder is decreased, the more easily the 100-mile column pushes air into the cylinder.

If the engine in question were to rise in the column of air, say for instance in a drive to the top of Pikes Peak, the pressure forcing its way into the engine would be lower. The cylinders would not contain as much air when the intake valve closes, the pressure in the combustion chamber would not rise as high as in Galveston, the pressure after ignition would not be as high, and less power would be produced.

Power comes from air. The earliest sailors of the sea knew this, every performance engine builder knows this, yet many people think that power comes from increasing the amount of fuel. Increasing the amount of fuel to increase the amount of power is true only to the extent that the extra fuel can add extra heat to the air to cause the air to expand. Add extra air

Using the air for power to move a vehicle is not a new concept to mankind. For centuries, mankind has known the great power potential of air.

The power of heated air has been understood only in the past few centuries. When air is heated, it expands rapidly in all directions. This rapidly expanding air, properly confined and directed, can be used to propel anything. Pressure, confinement, and direction of the expanding air can enhance the power created by heating the air.

and you add extra power. This book is the story of how to get extra air into the engine. The internal-combustion, spark-ignition engine works as a unit to pull in air. The compression ratio is a factor in how much the pressure is reduced when the piston moves down. The design of the ports in the cylinder head can affect how much air can flow into the engine. The specific area that this book will address is how to make the maximum amount of air possible available to the cylinder heads and how that air, once burned, will exit the cylinder head. This book is about the intake and exhaust systems.

Air temperature

Air temperature has a large effect on the amount of air entering the engine. Although the temperature will not have an effect on the volume of air entering the engine, it will have an effect on the amount of air molecules contained in a given volume of air. As the temperature of the air rises, the density of the air decreases. As the temperature of the air decreases, the density of the air increases.

Have you ever wondered why the little land rocket of your teens seemed to run a little better once the sun went down? Was this because of the excitement of a teenage night? After the sun goes down, the air cools, becomes denser, every cubic foot of air entering the engine contains more molecules. The more molecules of air there are entering the engine, the more potential power there is to be created from that engine.

Many fuel-injected engines use a mass air flow sensor to measure the total mass of air entering the engine. The engine monitors this input as grams per second of air flow. At sea level (29.92in of mercury pressure) and at 32 degrees Fahrenheit (F), 1,293 grams of air occupies 1

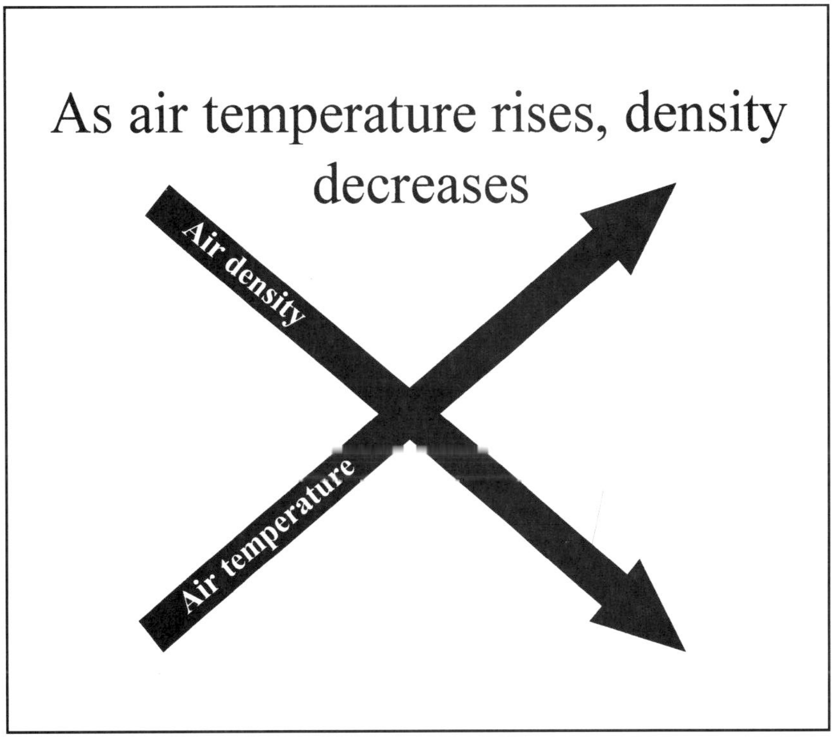

Air temperature can have a tremendous effect on the potential power of the air inhaled by the engine. Hot air has less density than cool air. Therefore, its ability to expand to pressurize the combustion chamber is decreased. The ideal power location for internal-combustion, spark-ignition engines is Death Valley, on a really cold day. This prospect not being realistic, try winter in Fairbanks, Alaska, where the altitude is only a few hundred feet and the air temperature can be a really dense -40 degrees F for days at a time.

Many fuel-injection systems use a mass air flow sensor to keep track of the volume, pressure, and density of the air moving into the intake system. These sensors provide information about the mass of air entering the computer so that the fuel-injection computer can adjust the fuel flow to compensate for the variation in the air characteristics.

Carburetors enjoyed many decades as the premier method of metering the fuel into the engine by adjusting the fuel flow rate to match the air flow rate and conditions. Carburetors are still the leading method of fueling a performance engine. The way they measure the characteristics inherently restricts air flow, and that limits power.

Air Density and Temperature

$$\frac{X}{1.37} = \frac{492}{560}$$

$$X = 1.37 \times \frac{492}{560}$$

$$X = \frac{674}{560}$$

$$X = 1.20$$

Density decreases inversely from the temperature.

When the air density decreases, the potential power of the engine decreases proportionally. The calculation above shows that when the temperature of the air entering the engine rises from 32 to 100 degrees F, the weight of a cubic foot of air decreases from 1.37oz to 1.20oz. Assuming the engine was running at 600cfm at both temperatures, the estimated power of the engine would decrease by the change in the air mass or (as discussed here) weight. The drop in potential power from the engine would be 12.5 percent.

cubic meter. This means that every cubic foot of air available to enter the engine has a mass of about 1.37 ounces (oz). If the engine is drawing 600 cubic feet per minute (cfm), the mass of the air entering the engine is 822oz per minute.

As the temperature increases, the air becomes even less dense. At 100 degrees F at sea level, the density will be less than at 32 degrees. To illustrate this fact we need to convert our temperature readings into Rankine. Rankine is temperature measured above absolute zero. Since 0 degrees F is 460 degrees Rankine (R), 32 degrees F is 492 degrees R and 100 degrees F is 560 degrees R. Density decreases inversely with the temperature. At 492 degrees R, there is 1.37oz of air per cubic foot. Therefore, 492 is to 560 as 1.37 is inversely to X.

$$\frac{X}{1.37} = \frac{492}{560}$$

$$X = 1.37 \times \frac{492}{560}$$

$$X = \frac{674}{560}$$

$$X = 1.20$$

When the temperature has risen from 32 degrees F to 100 degrees, the mass in a cubic foot of air has decreased from 1.37oz to 1.20oz. This means that the engine running at a speed where it draws 600cfm will now only be gulping 720oz of air instead of 822oz. This also means a corresponding potential horsepower loss of 12.5 percent. A well-designed intake manifold will minimize this power loss by reducing the pressure drop as the air flows through the manifold as much as possible.

Air pressure

When most people hear the topic of air density discussed,

they confuse it with pressure. In reality, pressure is one of the things that affects the density of a gas like air, but is not the only thing. Air pressure at the oceanic surface of planet Earth is 14.7psi, 29.92in of mercury or 101.3 kilopascals. At the bottom of the Dead Sea the pressure is about 15.28psi. This is an increase in pressure of about 3.8 percent. This means that the force trying to move the air into the combustion chamber is greater near the surface of the Dead Sea (1,310 feet (ft) below sea level) than it is at average mean sea level for the rest of the planet.

According to *The Automotive Handbook*, published by Robert Bosch Corporation, it is generally accepted in automotive engineering circles that a 1 percent change in power relates inversely to a 328ft change in altitude. This means that, theoretically, a 300hp engine will produce 304hp.

Now let us follow Bobby Unser to the top of Pikes Peak. The altitude there is 14,110ft. The atmospheric pressure will be a mere 8.62psi. According to the "328" rule, the 300hp engine will have only 57 percent of its original power. The 300hp engine will now be a 171hp engine. One of my favorite cars, the 40hp Volkswagen, will be a 22.8hp Volkswagen.

This is all well and good, but how does it get the car down the road? What does all this have to do with pistons, rods, and crankshafts? To create the maximum amount of power, the engine needs to get the maximum amount of air. The intake manifold restricts and limits the amount of air an engine can get. This problem is amplified when the atmospheric conditions are less than ideal. When the atmospheric conditions are less than ideal, the intake manifold and exhaust system must be as close to ideal as possible.

Humidity

Humidity is the least understood of the atmospheric conditions that affect air, air flow, and power. I once participated in a train-the-trainer where there was a very heated discussion of the effects of humidity on air density. The instructor was a fellow from Indianapolis who was determined that the moisture made the air heavier. A true statement, perhaps, but it misses the point. The constituent of the air we are concerned about is the oxygen content. *The Automotive Handbook* states, "Damp air contains less oxygen than dry air and thus reduces engine power. This power drop is generally negligible. The humidity of the damp, hot air in the tropics can noticeably reduce engine power."

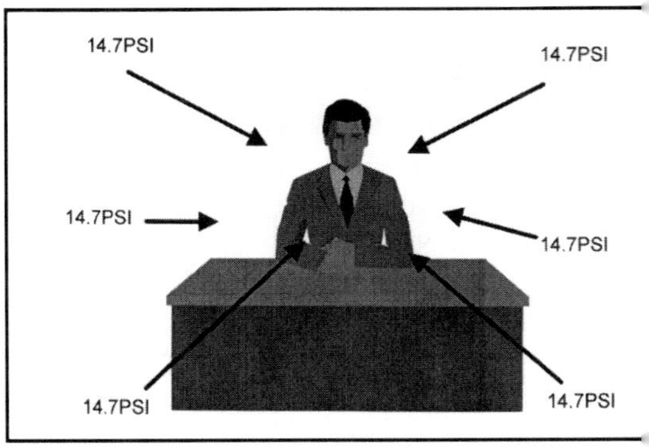

The man in this drawing is probably sitting at sea level. He is probably experiencing a "normal" day. On this "normal" day, the pressure around the man would be 14.7psi. When the weather or altitude changes, these changes will often be accompanied by a change in pressure. A rule of thumb is that the power potential of an engine will change inversely by 1 percent every time the pressure altitude changes by 328ft. Go up 328ft, the power potential decreases by 1 percent.

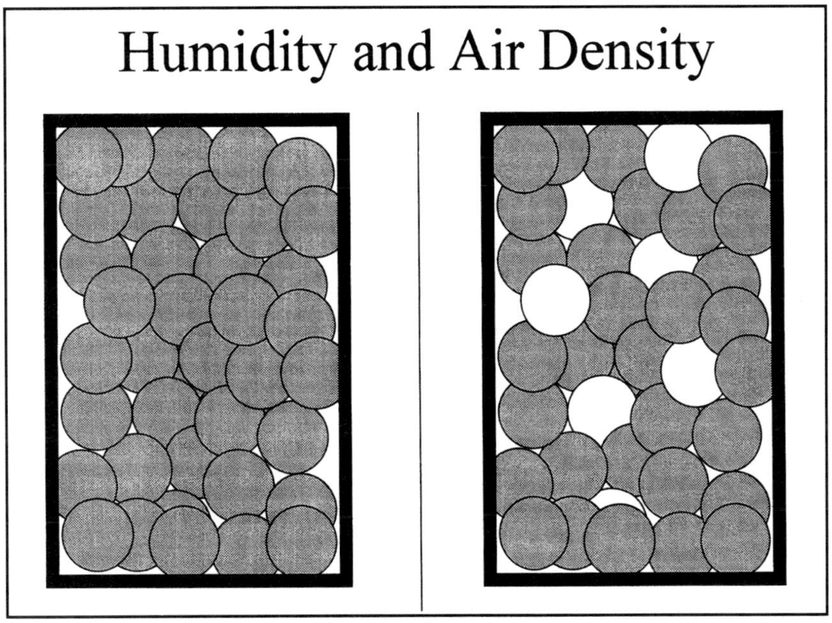

The effect of humidity on engine performance is often misunderstood. Many performance engine buffs believe that more humid air increases engine power. This misconception is the result of two factors that often accompany a spontaneous increase in humidity. The first cause of the misconception is that when it suddenly begins to rain, performance engines run better. The improvement in performance is not the result of an increase in humidity but rather a result of the fact that the rain is often accompanied by a decrease in air temperature; therefore, the air gets denser. Additionally, the cooler air decreases the likelihood of detonation. The humid air is less dense; this decreases the likelihood of detonation. The fact, however, remains: More humidity means less dense air, less dense air means less potential power.

Relative humidity is a familiar term to most people. It is a term you hear many nights on the local weather broadcast. Relative humidity is a term that describes the amount of water vapor there is in the air. When the temperature of the air increases by 20 degrees F, its ability to hold moisture doubles. This means that when the temperature rises from 60 to 100 degrees F, the amount of moisture the air can hold increases by six times. This is why, on an extremely hot, humid day, the potential moisture in the air can affect the power of the engine.

Air mass per unit of volume

Air mass per unit of volume is a function of air temperature, air pressure, and humidity. All of these things affect the air mass in each cubic foot of air entering the engine. The ideal situation would be 0 percent relative humidity, -40 degrees F, and on a drag strip near the Dead Sea. Under these conditions you could get maximum power from your engine. Everything that alters these "ideal" conditions reduces the potential power of the engine.

If the day is hot, the air is less dense and there will be less expandable air molecules in the air mass entering the engine. Power is reduced. In the summer of 1989 it was so hot in Phoenix, Arizona, that aircraft jets could not create enough thrust to get the planes off the ground. Atmospheric pressure reduces power at the rate of 1 percent every 328ft. Humidity reduces power. The higher the relative humidity, the less power potential there is from the engine. The actual amount of moisture in the air at a given relative humidity doubles for every 20 degrees F the temperature increases. Therefore, the effects of humidity are far more noticeable when the humidity is high on a hot day.

The bottom line is: avoid racing on the top of Pikes Peak when the temperature is 90 degrees F and the relative humidity is 99 percent.

Engine air requirements

Theoretical air flow

The amount of air drawn in by the engine increases as the size and speed of the engine increase. Here is the formula for theoretical air flow:

$$\text{theoretical airflow} = \frac{\text{rpm x displacement}}{3456}$$

Using this formula, we find the rat motor I dream about putting into my Volvo 544 would have a theoretical airflow of:

$$\text{theoretical airflow} = \frac{4000 \times 454}{3456}$$

$$\text{theoretical airflow} = \frac{1{,}816{,}000}{3456}$$

$$\text{theoretical airflow} = 525.63\text{cfm}$$

This would actually seem disappointing for those who just laid out $500 or more for a Holley 1100 to put on top of their 454.

The job of the intake system is to ensure the engine can draw in from the atmosphere all the air it could demand under any operating conditions and at any speed. The job of the exhaust system is to ensure that the engine can push out all the waste gasses created through the process of combustion, no matter what the engine operating conditions or engine speed might be.

Theoretical air flow assumes that every engine has a 100 percent volumetric efficiency. Without turbocharging or supercharging, no engine enjoys 100 percent volumetric efficiency. In the ideal world, each cylinder of an internal-combustion engine would be like a gentleman's belt. When a man buys a new belt, a belt selected so there will be a couple of unused notches near the end of the belt, one law of physics and the universe always comes into play.

Physical law of the universe #1: The size of a man's waist will always expand so that the last notch of his longest belt will be used. Thus a man's waist, with respect to his belts, shall always achieve 100 percent volumetric efficiency.

The cylinders of the internal-combustion engine are not at all like this. Most engines never achieve 100 percent volumetric efficiency; those that do achieve it only in a very narrow rpm band.

Actual air flow

The actual air flow into an engine and its volumetric efficiency are impossible to determine mathematically; they must be determined through experimentation and measurement. Many modern engines have good volumetric efficiency. The General Motors Quad 4 engine, for example, makes use of excellent scavenging characteristics of the exhaust system to "supercharge" the intake. As the exhaust gasses flow out of the combustion chamber, they create a lower pressure in the combustion chamber than would be developed by the piston pushing the gasses out alone. This helps to draw the intake air charge into the combustion chamber. Additionally, the flow of the exhaust gasses helps to create an inertial movement of the intake gasses, which further helps to draw them into the combustion chambers. Because the intake and exhaust systems are so well tuned to one another, the engine achieves volumetric efficiency at several engine speeds.

Generally, it is accepted that in practice an engine designed and built for street operation will have a volumetric efficiency of 85 percent. On the other hand, an engine built for racing use, but not equipped with a supercharger, will have a volumetric efficiency of 110 percent.

$$\text{street cfm} = \frac{\text{rpm x displacement}}{3456} \times 0.85$$

$$\text{street cfm} = \frac{4000 \times 454}{3456} \times 0.85$$

$$\text{street cfm} = \frac{1,816,000}{3456} \times 0.85$$

$$\text{street cfm} = 525 \times 0.85$$

$$\text{street cfm} = 446$$

$$\text{racing cfm} = \frac{\text{rpm x displacement}}{3456} \times 1.0$$

$$\text{racing cfm} = \frac{4000 \times 454}{3456} \times 1.1$$

$$\text{racing cfm} = \frac{1,816,000}{3456} \times 1.1$$

$$\text{racing cfm} = 525 \times 1.1$$

$$\text{racing cfm} = 578$$

Cubic feet per minute air flow is essential to provide the power we demand from the engine.

To carry the cfm requirement to a negative extreme, I was once working as a technician in an import car repair shop near Seattle, Washington. We had a customer come in with the complaint that his Subaru had a top speed of less than 30 miles per hour (mph). The car had already been into several dealerships and independent shops in the area. An amazing list of what had been done was quoted to me. These things included rebuilding the carburetor, replacing the spark plugs, replacing the fuel filter, and even servicing the transmission. The problem was Mount St. Helens volcanic ash clogging the air filter. Without the air flow, there was no power; without the power, the top speed was extremely low. Air flow is power.

Theoretical air flow

$$\textit{theoretical airflow} = \frac{rpm \times displacement}{3456}$$

$$\textit{theoretical airflow} = \frac{4000 \times 454}{3456}$$

$$\textit{theoretical airflow} = \frac{1,816,000}{3456}$$

$$\textit{theoretical airflow} = 525.63\ cfm$$

The theoretical air flow of an engine is a mathematical calculation based on the swept volume of the combustion chambers and the speed of the engine. In the above example, the speed of the engine is 4000rpm and the size of the engine is 454ci. The theoretical air flow is therefore 525.63cfm.

Street CFM vs. Racing CFM

$$street\,cfm = \frac{rpm \times displacement}{3456} \times 0.85$$

$$street\,cfm = \frac{4000 \times 454}{3456} \times 0.85$$

$$street\,cfm = \frac{1,816,000}{3456} \times 0.85$$

$$street\,cfm = 525 \times 0.85$$

$$street\,cfm = 446$$

Estimated actual street CFM

Estimated actual racing CFM

$$racing\,cfm = \frac{rpm \times displacement}{3456} \times 1.$$

$$racing\,cfm = \frac{4000 \times 454}{3456} \times 1.1$$

$$racing\,cfm = \frac{1,816,000}{3456} \times 1.1$$

$$racing\,cfm = 525 \times 1.1$$

$$racing\,cfm = 578$$

In the absence of sophisticated measurement equipment, actual cfm can only be estimated. There are two common formulas for calculating actual air flow. Both formulas are an estimate based on theoretical cfm. The "street cfm" is an estimate of how much air the typical nonracing application might consume at a given engine speed. The theoretical cfm is multiplied by a factor of 0.85. This assumes that the volumetric efficiency of the engine is 85 percent at the speed for which the calculation is being done. For "racing cfm" the assumed volumetric efficiency is 110 percent; therefore, the theoretical cfm should be multiplied by a 1.1 factor.

Reducing pressure in the intake system (vacuum)

The dropping piston creates a low-pressure area

In most, though not all, engines the pistons travel in pairs. The piston pairs are known as companion cylinders. In most four-cylinder engines when piston number one is in the power stroke, piston number four is on the intake stroke. When cylinder number one fires, the expanding gasses force the piston down the cylinder. As piston number one is forced down the cylinder, piston number four is being carried along with it. Piston number four is creating a low-pressure area in its cylinder. Providing the intake valve is open, which it should be, the cylinder will now fill with air. To ensure that the drop in pressure caused by the retreating piston occurs, it is necessary that there be good cylinder integrity. A leaking head gasket, worn rings, or a poorly sealing valve can cause the low pressure created by the piston to pull air or other substances into the cylinder. To ensure that the cylinder fills properly with measured air, it is necessary that the only path for the air into the cylinder be through the central area of the intake manifold.

It is really too bad that Venus does not have a 21 percent oxygen atmosphere like planet Earth. In a typical internal-combustion engine we are lucky to reduce the pressure in the intake system to one-third that of atmospheric. The action of the pistons in the cylinder, therefore, only creates a pressure differential (vacuum) of 10psi (20in of mercury) at sea level when the engine is idling. As the altitude increases and atmospheric pressure drops, the difference between atmospheric pressure and intake manifold pressure at an idle remains at a relatively constant 1:3 ratio. When you cross over Donner Pass in California (altitude 7,200ft), the atmospheric pressure is about 11.37psi (about 23in of mercury). The pressure in the intake manifold is about 3.8psi (about 7.7in of mercury). These are pressure readings. The typical motorhead is more familiar with vacuum readings. If the atmospheric pressure is 23in of mercury and the pressure in the manifold is 7.7in of mercury, then:

vacuum = atmospheric pressure - manifold pressure
vacuum = 23 - 7.7
vacuum = 15.3

An engine that produces a pressure differential (vacuum) of 20in of mercury at sea level will only produce a pressure differential of 15.3in of mercury at Donner Summit. The result of this loss of a pressure differential is a loss of force to push the air into the cylinder as the piston retreats. With less force, less air enters the cylinder, and there is a lower potential for power.

On a more motorhead-friendly planet, such as Venus with a 21 percent oxygen atmosphere, the atmospheric pressure would be higher. On Venus the atmospheric pressure is 1,352psi, ninety times that of planet Earth. That means that the estimated pressure differential would be 892psi. Wow! That would be like running 877psi of boost on your turbocharger. Can you imagine a 40hp Volkswagen that could turn low 5s in the quarter mile? Of course there is the question of aerodynamics in that thick atmosphere, but that is a topic for another book.

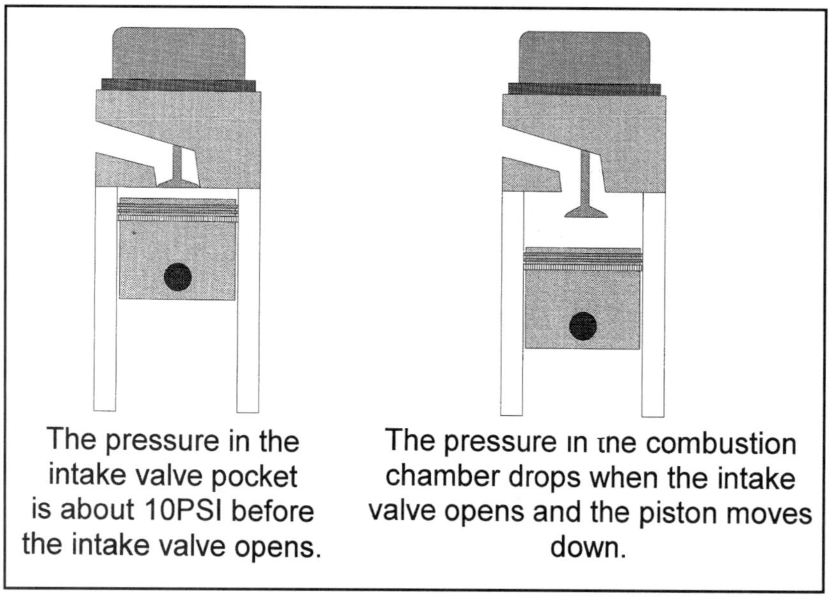

The pressure in the valve pocket of the cylinder head is typically only about 10psi just prior to the opening of the intake valve. The pressure drops when the intake valve opens and the piston begins to move down. The low pressure area in the cylinder causes the higher pressure air of the atmosphere to move in to fill the void. If the pressure in the cylinder when the piston reaches bottom dead center is greater than atmospheric pressure, then we can say that the cylinder has been charged to greater than 100 percent volumetric efficiency.

Valve action

The four-stroke engine common to the automobile today was the brainchild of nineteenth-century engineer N.A. Otto or nineteenth-century French engineer De Rochas, depending on which

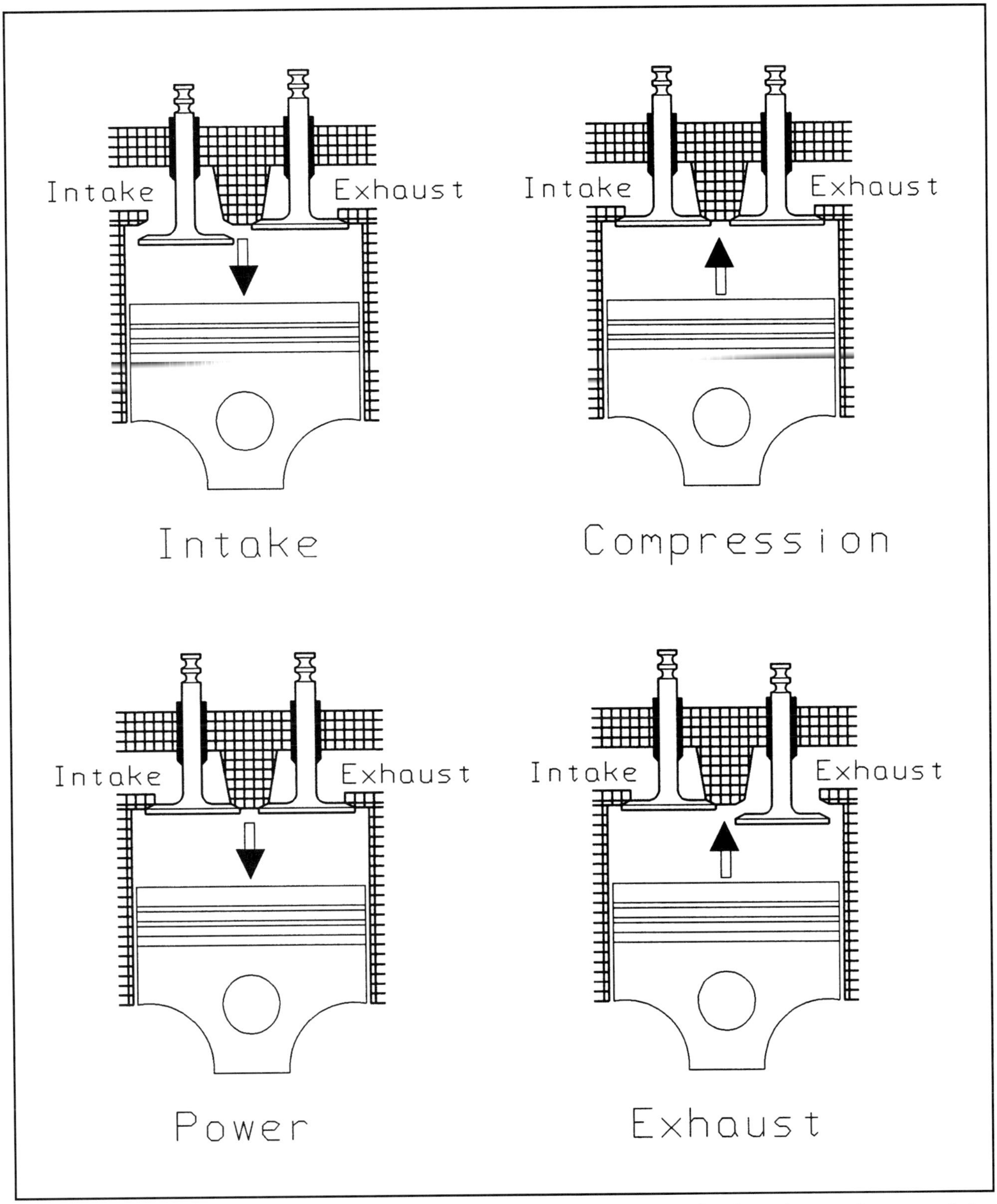

Most modern gasoline spark-ignition engines are four-stroke engines. This technology dates from the mid-1800s. In these engines, air and fuel are drawn into the cylinder as the piston moves down during the intake stroke. During the compression stroke, the piston squeezes the air/fuel charge into a dense mixture. A spark ignites the fuel, causing it to expand rapidly and also heat up the air in the cylinder. The expanding air and fuel drives the piston down on the power stroke. After reaching bottom dead center, the exhaust valve opens and the spent air/fuel charge is pushed out of the cylinder by the rising piston.

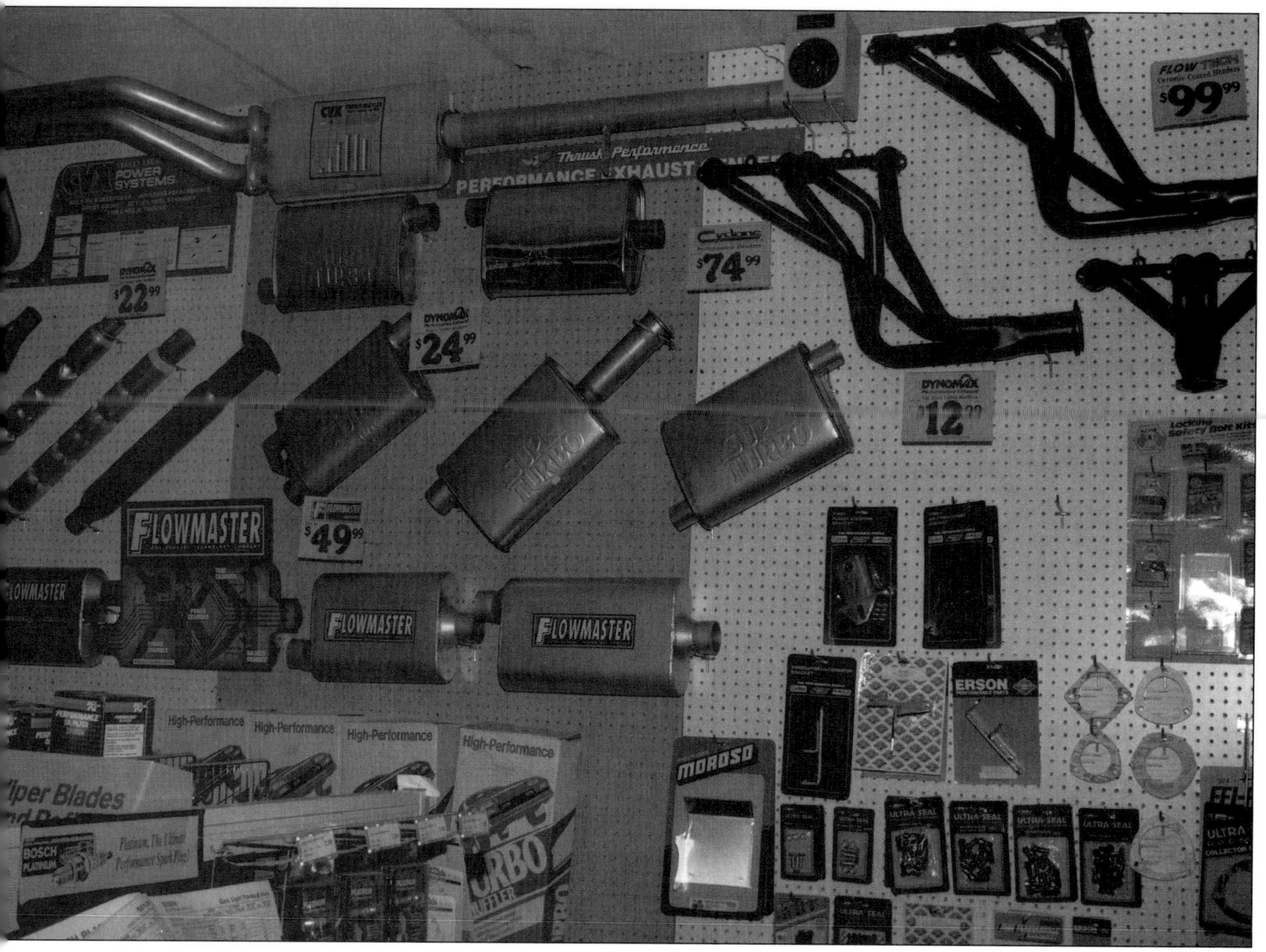

In the ideal world, there would be a positive method of extracting the exhaust gases from the cylinder. Something close to this can be achieved with the right exhaust system. At the point where multiple exhaust pipes join together, the flow of exhaust gases out of pipe A creates a low-pressure area in pipe B. This low-pressure area provides an increased pressure differential for cylinder B to move its exhaust gases into when the exhaust valve for B opens. This process is called scavenging.

side you take in the 1876 lawsuit between these two gentlemen. The four strokes used in this type of engine are intake, compression, ignition, and exhaust. I like to think of them as suck, squeeze, boom, belch. The opening and closing of the valves is controlled by the camshaft and must be precisely synchronized to the rotation of the crankshaft. On the intake stroke, the intake valve must be open and be the sole path for air to enter the cylinder. On the compression stroke, the valves must seal the air/fuel charge. This is important for the exhaust valve, but it is even more important for the intake valve. Not only will power be lost due to the expanding gasses of combustion not being held in the cylinder, but if it leaks through the intake valve it will also pressurize the intake manifold. When the intake manifold becomes pressurized, the pressure differential between the manifold and the atmosphere reduces. This causes the air flow into the manifold and therefore into the engine to decrease.

Scavenging

The dictionary defines scavenging as: 1. To remove filth, rubbish and refuse from . . . 2. To remove exhaust gasses from. . . . Both of these definitions are appropriate. This topic has to do with the efficiency with which the exhaust system cleans the cylinder before the intake system begins to refill the cylinder.

The exhaust valve opens just prior to the piston reaching bottom dead center (BDC). At this point, the gasses in the combustion chamber are still at an ex-

Not all forms of supercharging include huge mechanical devices. A passive system such as intake manifold design can also improve air flow into the cylinders, and at a few engine speeds and throttle positions it can produce a positive pressure in the intake manifold.

tremely high pressure. It is during this time that almost 50 percent of the combusted gasses leave the combustion chamber, forced out by their own pressure. As the piston travels up during the exhaust stroke, it pushes most of the remaining gasses out of the cylinder. When the piston approaches top dead center (TDC), the intake valve opens. By this time the exhaust gasses have a considerable momentum; at this point both the intake and the exhaust valves are open. This is called valve overlap or gas exchange TDC. The velocity of the exiting exhaust gasses not only creates an inertia that moves the exhaust gasses out of the cylinder but also helps to draw in the intake air charge.

Pressurizing the intake system

Dynamic supercharging

When the average person talks about supercharging, he or she imagines a big blower sitting on top of the engine, or Don Garlits (is my age showing?) sitting behind a monster "gimmy" blower with flames shooting out of the exhaust (yes, my age is definitely reflected in this). In reality, dynamic supercharging involves a much more subtle approach. Dynamic supercharging maximizes the use of the 100-mile column of air to force the air into the cylinders. There are two types of dynamic supercharging: ram pipe supercharging and tuned-intake tube charging.

Ram pipe supercharging

In this technology, each cylinder has its own intake manifold usually connected to a common receiver. This design uses the kinetic energy of the gas moving into the cylinder to press the air into the cylinder. In a sense, it uses the air as a "mallet" to pound the air into the cylinder.

Tuned-intake tube charging

This intake system is just like the ram pipe system, except the intake runners are grouped together by ignition intervals.

Turbocharging

Turbocharging is a fascinating technology and will be covered thoroughly later in the book. Essentially, a turbocharger takes energy that would have gone out

Both engine-driven mechanical superchargers (like this) and exhaust-driven turbochargers can greatly increase an engine's performance. By forcing air into the cylinder until the pressure in the cylinder is greater than atmospheric, they can increase the effective compression ratio. Properly chosen, tuned, and installed, these systems can raise "compression ratio" at the points where performance needs to be improved and leave it the same under all other engine operating conditions.

the tailpipe as heat and converts it into energy. The energy then drives a turbine that spins a compressor. The compressor increases the pressure in the intake manifold. The air molecules become squashed together, cramming more oxygen molecules into the cylinders. Since more air in the cylinders translates directly as more power from the engine, this is an efficient way to increase the power output of an engine. Some turbocharged engines run over 100psi of boost. This means that the mass of air in the cylinders can be more than six times greater than without a turbocharger. The power potential of the engine is many times greater than without the turbocharger. Of course, all good things do have a downside. When the boost pressures are this high, the engines have a short life expectancy, perhaps seconds.

Supercharging

A supercharger is an engine-driven pump used to force air into the cylinders. This is a topic that will be covered later in this book as well. Engine-driven superchargers are what people typically think of when the subject of superchargers comes up.

Air-fuel ratio

Theoretical correct air-fuel ratio (stoichiometry)

Gasoline is a complex hydrocarbon made up of approximately 86 percent carbon and 14 percent hydrogen by weight. There are trace impurities that contribute to the noxious soup from the tailpipe. Sulfur can combine with oxygen during the combustion process, producing sulfuric acid and sulfur dioxide. There are also a wide range of additives that are used.

Until recently, tetraethyl lead or tetramethyl lead were used as knock inhibitors. When these oxidized in the combustion chamber, they not only produced an acid but also contributed to the background lead contamination of the environment.

As odd as it may sound, the air itself is one of its own prime defilers. About 78 percent of the air we breath is nitrogen. When this nitrogen is combined with oxygen during the combustion process, it can produce both nitrogen monoxide and nitrogen dioxide.

Air is made up of 78 percent nitrogen and 21 percent oxygen. The remaining 1 percent consists of various trace gases such as xenon, neon, and argon.

The perfect mix of fuel and air are 14.7 parts of air by weight to one part of fuel by weight. When mixed in these proportions, theoretically there will be no air and no fuel left at the end of combustion. In fact, in theory there will be only carbon dioxide (CO_2) and water. We will see later that this is not reality.

Keeping the charge in suspension

When it comes to mixing fuel and air, there are two basic types of intake manifolds, wet and dry. Actually, for a gasoline spark-ignition engine the concept of a dry manifold is not proper. Even in these manifolds the air and fuel are mixed just above the intake valve. Only direct-injected engines, engines where the fuel is injected directly into the combustion chamber, are true dry manifold engines.

This may be where the real trick lies. Wet manifold engines use carburetors or throttle body injection to mix the fuel with the air destined for the combustion chambers. As the air/fuel charge moves through the intake runners, there is a tendency for some of the fuel to drop out of suspension. Imagine a mountain stream for a moment. As the rushing waters come to

This drawing is a two-dimensional graphic used to illustrate a three-dimensional concept. The correct air/fuel ratio is 10,000 volume units of air to 1 volume unit of fuel. Since air is far less dense than fuel, this equates to an air/fuel ratio of about 14.7:1 by weight.

level ground, they slow; as they slow, rocks begin to drop from the moving water. The slower the waters move, the smaller the rocks that will drop out of suspension. The same thing happens in an intake manifold. Fuel particles will drop out of suspension as the air flow through the intake slows. This dropout is held to a minimum on multipoint injected engines. These "dry manifold" engines mix the fuel with the flowing air only just above the intake valve. As a result, the fuel is more evenly and accurately carried to the cylinder.

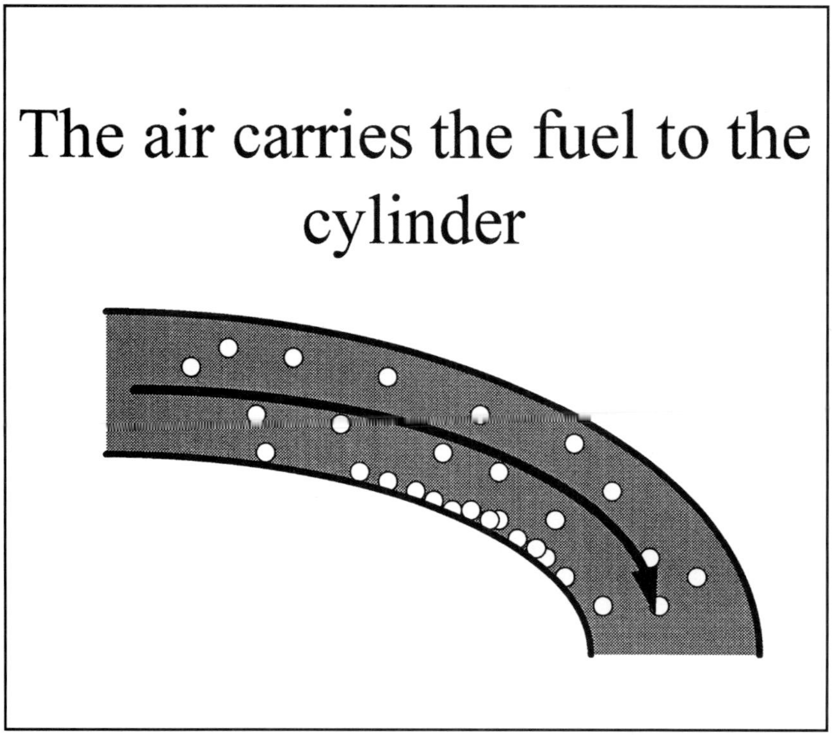

On its way to the combustion chamber, the air has another task to perform. It is the primary agent for transporting the fuel to the combustion chamber. Since most engines destined for performance modification today are still carbureted, inadequate air flow characteristics can result in poor air/fuel distribution and poor fuel mixture quality in the combustion chamber. If this happens, it makes no difference how well designed the engine is; performance will suffer.

Chapter 1
Tools

In the early 1990s, the first television show to ever explain a man's relationship with his tools came on the air. The name of the show was *Home Improvement*. Weekly this show explored the peculiar relationship that men have with tools. There is little doubt that it was a prehistoric man who first picked up a stick to use for getting grubs out of a rotting tree trunk. If it had been left up to women, we would all still be eating slightly rotting vegetables that had fallen to the ground.

Of course, another theory holds that women invented tools just to make sure that prehistoric man always had a "honey-do" list around the cave. The really depressing part of this theory is that it requires prehistoric man to have been stupid enough to go along with it and even develop a fondness for it. No, no, no . . . it can't be.

Compression tester

The compression tester is a pressure gauge with a check valve. When installed in a spark plug hole, the compression gauge is able to record and display the amount of pressure created in the cylinder as the engine is cranked.

To get ready to run a compression test, remove all of the spark plugs, disconnect the primary power supply to the ignition coil, and block the throttle wide open. Install the compression gauge in the number one spark plug hole. Crank the engine twelve revolutions. Listen to and count the "puffing" noises. Record the compression reading and move to the number two cylinder. Repeat the process on each of the cylinders. When done, compare the compression readings. If any of the cylinders is considerably different than the average of the rest, repeat the test. When you repeat the test, squirt about a teaspoon of oil into each cylinder before testing it. This is called a wet compression test.

Cylinder leakage tester

The cylinder leakage test is superior to the compression test, although its goals are the same. It also requires more equipment. With the piston at TDC compression stroke, connect the cylinder leakage tester to cylinder number one. Measure the percentage of leakage. Additionally, use a piece of heater hose held to the ear to determine where the leakage is. Stick the heater hose in the open bore of the fuel-injection throttle assembly or carburetor. If there is a great deal of air es-

Measuring engine compression can give you a good idea of the condition of the cylinders and pistons before beginning your modifications. After completing your repair, overhaul, or modification, the compression should be checked. The compression should be even in all cylinders. Uneven compression after an overhaul can be indicative of improperly installed rings or valve adjustment problems.

A cylinder leakage tester can be used to pinpoint where the loss of integrity is in a cylinder. For instance, if after an overhaul the compression is low in one cylinder, set up the cylinder leakage tester on that cylinder. Bring the piston to top dead center on the compression stroke and use the tester to blow compressed air into the cylinder. The dial on the tester indicates the percentage of air going into the cylinder that is escaping from the cylinder. If the loss is greater than 10 percent, use a heater hose as a stethoscope to pinpoint the location of the leak. If the air is escaping through the intake or exhaust pipe, the loss is valve related. If the air is escaping into the radiator, then there is a head gasket problem. If you place the end of the hose into the oil filler and hear air escaping, then the problem is related to the pistons or the rings.

caping through here, you have an intake valve problem. Stick the hose in the exhaust; if there is a great deal of air escaping, you have a bad exhaust valve. Observe the radiator coolant. If there are a lot of bubbles in the coolant, you have a blown head gasket, cracked head, or cracked block. Next remove the oil filler cap. Place the heater hose in the oil filler neck or valve cover opening. If you hear excessive amounts of air, the rings are allowing the air to bypass. Now place the hose over the spark plug holes of the adjacent cylinders. If you hear a significant amount of air from one of the adjacent cylinders, then the head gasket is blown between the cylinder being tested and the cylinder where the air is heard.

You will hear air leaking in all of the places discussed above. The sound of the air will only be significant when the cylinder leakage indicated is excessive. The difficult part of the cylinder leakage test is determining how much leakage is too much. Opinions differ on this topic as much as they do on the Kennedy assassination. Manufacturers such as Honda say that over 5 percent leakage is too much. Some professional technicians feel that 15 percent is acceptable. The truth of what is acceptable lies somewhere in between. Being realistic, you have gone to the trouble of performing the cylinder leakage test because there is an obvious problem. Generally speaking, the types of problems that would be caused by 5 to 10 percent leakage would not be noticed. You are looking for a problem resulting from cylinder leakage that is probably greater than 15 percent.

Oil pressure gauge

The oil pressure gauge is a mechanical gauge that screws into one of the oil galleries in the engine block. When the engine is started, the oil pressure should rise to the proper specification or higher.

Piston ring pliers

One of the trickiest operations of rebuilding an engine is to remove and install the brittle piston rings without breaking them. While the rings can easily be spread far enough to go over the piston head and slip them into the ring grooves of the pistons, they can also be easily spread far enough to break them. While the piston ring pliers do not eliminate the possibility of breaking the piston rings, they do decrease it.

Torque wrench

The most essential tool for rebuilding engines is the torque wrench. All bolts and other fasteners stretch when they are tightened. If they are tightened too much, their threads will strip or they will shear. If they are not tightened enough, the bolt will loosen. Spare no expense when you purchase a torque wrench.

There are two types of torque wrenches available: the beam type and the click type. The beam type uses a pointer which indicates the amount of torque on a scale attached to a flexible bar. As torque is applied to the bolt, the flexible bar bends. The amount of bending in the bar indicates the amount of torque being applied to the bolt. The beam torque wrench was the industry standard for decades and in highly skilled hands is still an adequate tool. Its biggest weakness is its low consistency of accuracy, even in the most skilled hands.

The click torque wrench is far more accurate, especially for the occasional user. The desired torque is dialed in. When the

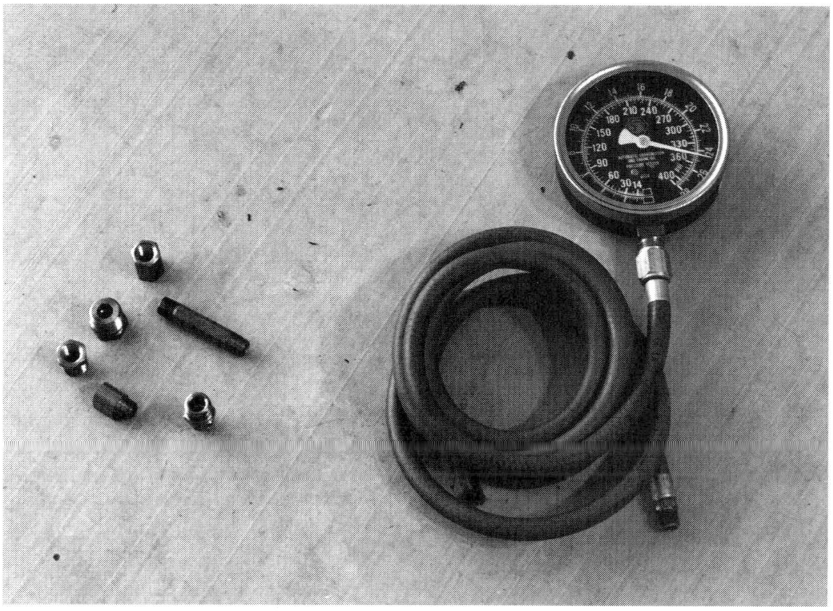

Can you imagine leaving on a thousand-mile journey in a car you have never seen before without taking at least a short test drive first? Can you imagine running an engine for an extended period of time without testing the oil pressure? Although oil pressure alone does not ensure that the engine is being properly lubricated at all critical points, it does provide a good indication.

Although piston rings can be installed without the use of specialty pliers, they cannot be replaced without at least a small risk of breaking one or more of them. Piston ring installation pliers virtually eliminate this risk for less than $10.

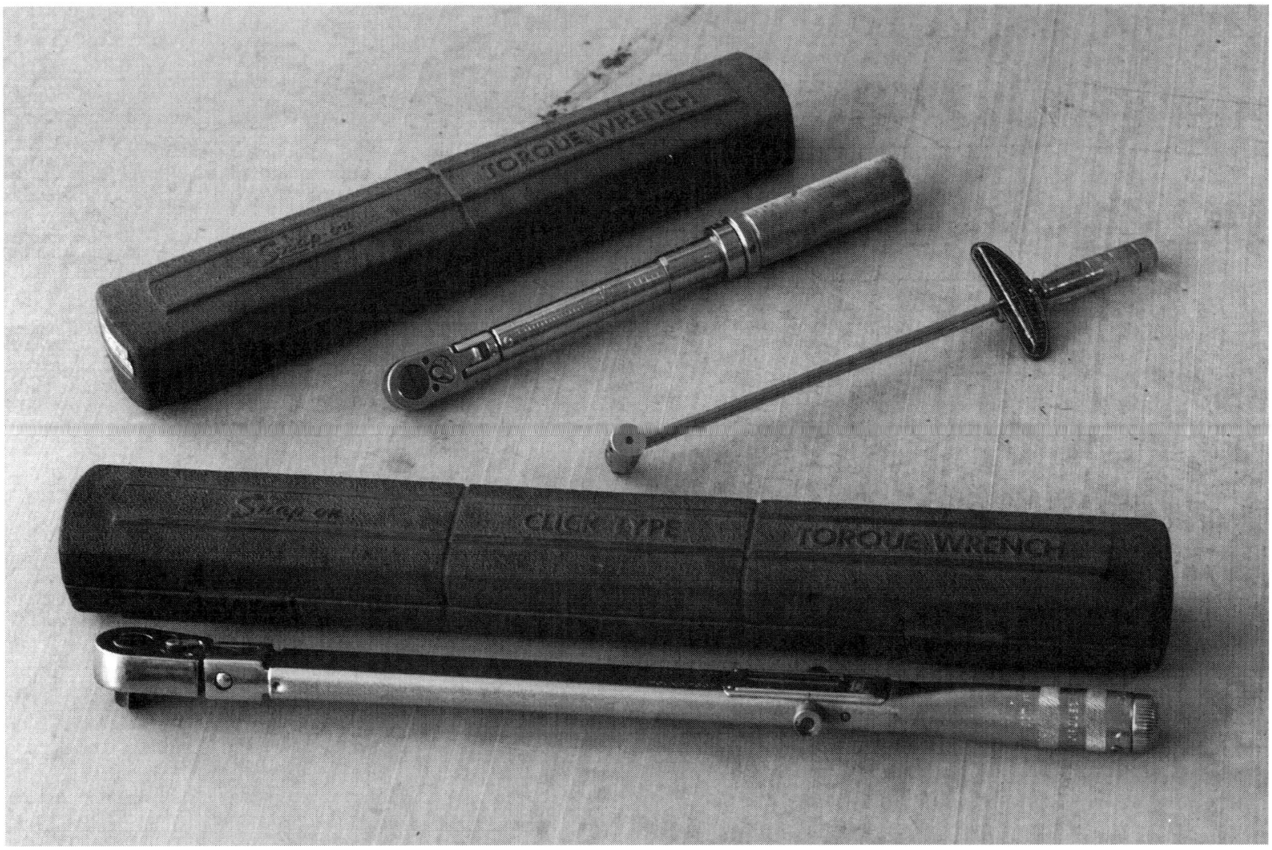

I once worked with a mechanic who was twice my age and swore he could overhaul an engine and never touch a torque wrench. Funny thing is, when I worked with him we were both tire mounters. Could there be a relationship? Anyway, a quality set of torque wrenches is indispensable in doing quality engine work. Spare no expense.

The piston ring compressor is essential for holding the piston rings deep in their grooves as the piston is installed. If a ring hangs up on the lip of the cylinder during installation, there is a risk of breaking the ring. There are many types of ring compressors on the market. I prefer this ring compressor. A simple lever action clamps the tool around the piston and releases it.

prescribed torque is reached, a click is heard and felt indicating to the user to stop tightening the bolt. These torque wrenches feature a high degree of accuracy. Although the click torque wrench can cost dozens of times more than its beam counterpart, it can be money well spent. One incorrectly torqued bolt can result in having to completely redo the rebuild, this is both expensive and time consuming.

Piston ring compressor

After the rings have been installed on the pistons, there is still the challenge of slipping the pistons into the cylinders in spite of the fact that the rings do not fit. Each ring is shaped so that it is larger than the cylinder in order to maintain firm contact between the ring and

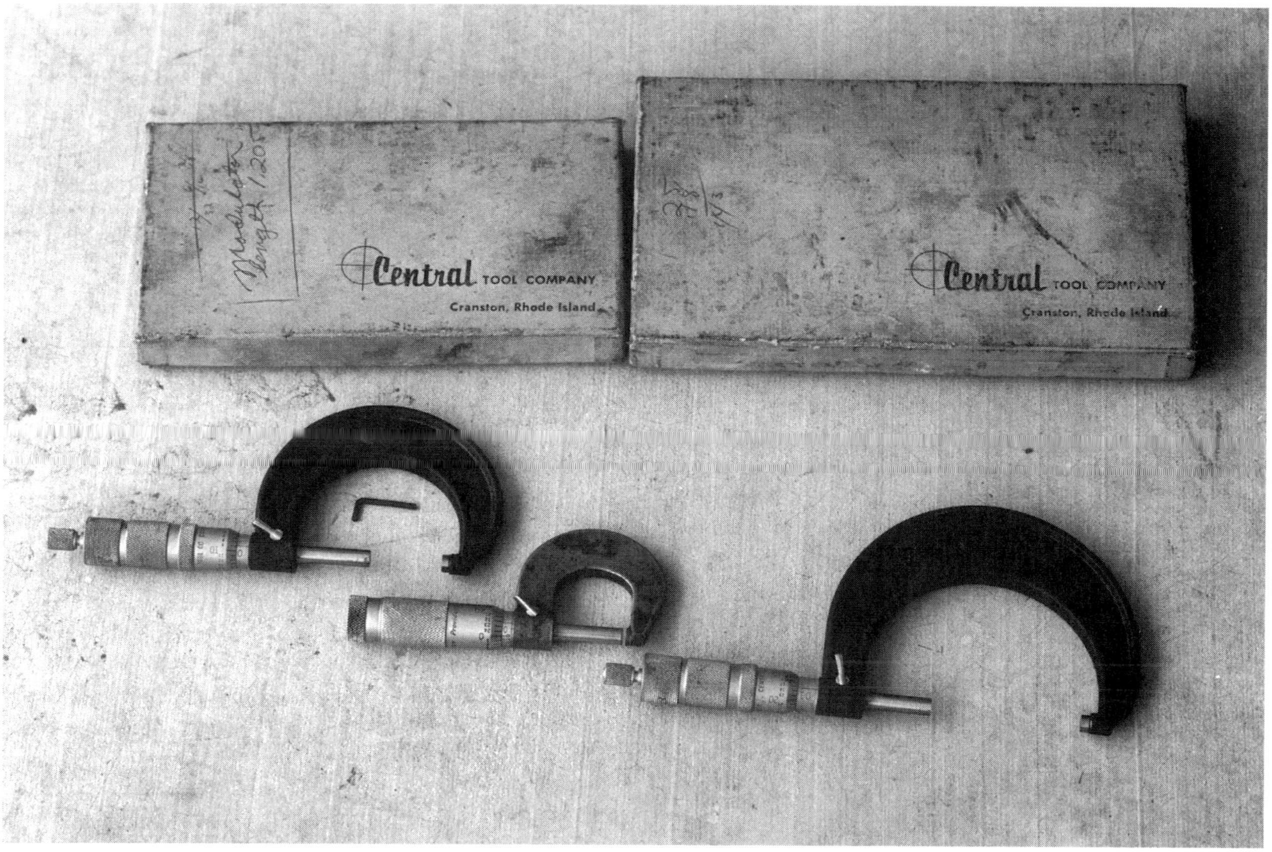

A good set of micrometers is absolutely essential to quality engine building. A lot of money can be invested in these tools. I rescued mine from a pawn shop for about half their new price. Be sure to check used micrometers carefully for accuracy. One drop by the former owner can damage them beyond use.

the cylinder wall. A gap in the ring allows it to be squeezed together so it will fit in the cylinder. The open end of the rings must be squeezed together to get the pistons into the cylinder. This is the job of the piston ring compressor.

While any piston ring compressor will work for most engines, some engines require a special compressor. With most domestic and foreign engines the pistons go in from the top. A few engines, such as the old Volkswagen air-cooled engines, require the pistons to be installed from the bottom. On these applications the connecting rod is attached to the crankshaft before it is installed into the cylinder. A standard ring compressor could not be removed once the piston is installed. Although there are ways around this problem, there is a special ring compressor available, one designed to come apart for easy removal.

Micrometer

An automobile engine is made up of dozens of closely machined parts. Measurements of these parts requires precision to the thousandths of an inch. The micrometer provides this level of accuracy. A basic set consists of a 0 to 1in, 1 to 2in, 2 to 3in, and 3 to 4in. When working with larger engines, big block and truck engines may require a 4 to 5in micrometer or larger. A precision instrument, micrometers should be handled with extreme care.

The micrometer is a simple device in spite of its high level of accuracy. The basic parts of it are the spindle, the thimble, the anvil, and the barrel. The spindle is threaded at a standard of forty threads per inch. Rotating the thimble moves the spindle 1/40 or 25/1,000 of an inch (0.025in). On the standard micrometer, there are twenty-five graduations around the thimble. This configuration allows an accuracy of 1/1,000in (0.001in). The barrel of the micrometer is marked off in 0.025in units. There is a numerical marking every four marks. The numerical marking, therefore, indicates a movement of 1/10in.

Using a micrometer involves placing its anvil and spindle against opposite ends of the object to be measured. Rotate the thimble until the spindle makes firm but not heavy contact with the object. Read the highest numerical marking visible on the barrel. This will indicate the

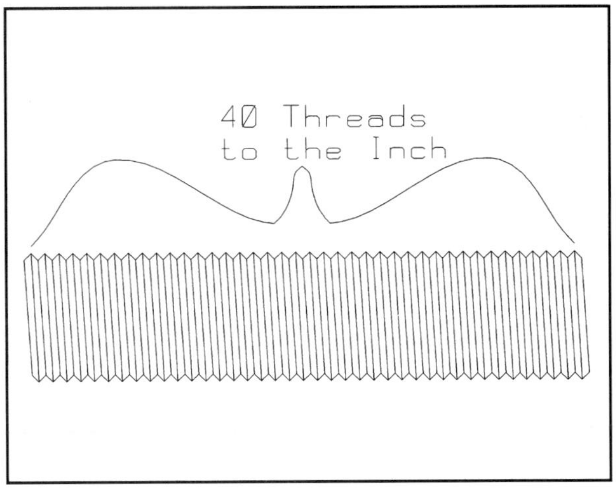

A micrometer is nothing more than a threaded device with exactly forty threads to the inch. Therefore, a single rotation of the barrel of the micrometer will move it 0.025in along the threads. Therefore, 1/25 of a rotation of the barrel equals 0.001in (1/1000in).

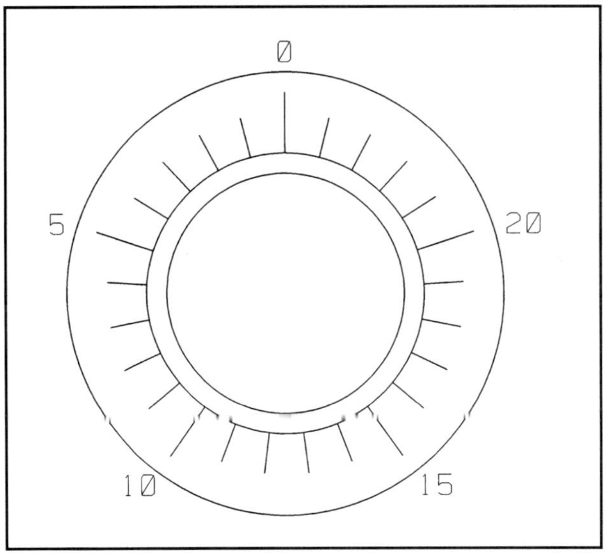

The barrel is marked off in twenty-five units of 0.001in. One full turn of the barrel represents 0.025in.

nearest tenth of an inch. Counting the barrel markings beyond the visible numerical mark will indicate the nearest 0.025in. Observe the twenty-five marks on the thimble and how they align with the line on the barrel; this will indicate the measurement to the nearest thousandth of an inch.

A quality micrometer should be treated as the precision instrument it is. Over-tightening when taking measurements can stretch the frame, decreasing the accuracy of the micrometer. Often this can be remedied by adjustment using a gauge block. Dropping the micrometer can damage it beyond adjustment. Top technicians may be more inclined to loan you their spouses than their micrometers.

Inside micrometer

The inside micrometer works just like the outside micrometer described above; however, its design allows it to be used to measure the inside of a bore.

Telescoping gauges and split ball gauges

When the cylinder or bore is smaller than 2in, the inside micrometer cannot be used. Extremely small bores require the use of split ball gauges. Inserted into the bore of something like a valve guide, the split ball can be adjusted to the diameter of the guide. When removed from the bore, a standard micrometer can be used to measure the size of the split. This will be the diameter of the guide.

Telescoping gauges are available in sizes up to several inches. They consist of spring-loaded rods that make contact with the sides of the cylinder or bearing bore to be measured. When removed, the length of the telescoping rods can be measured with a standard microme-

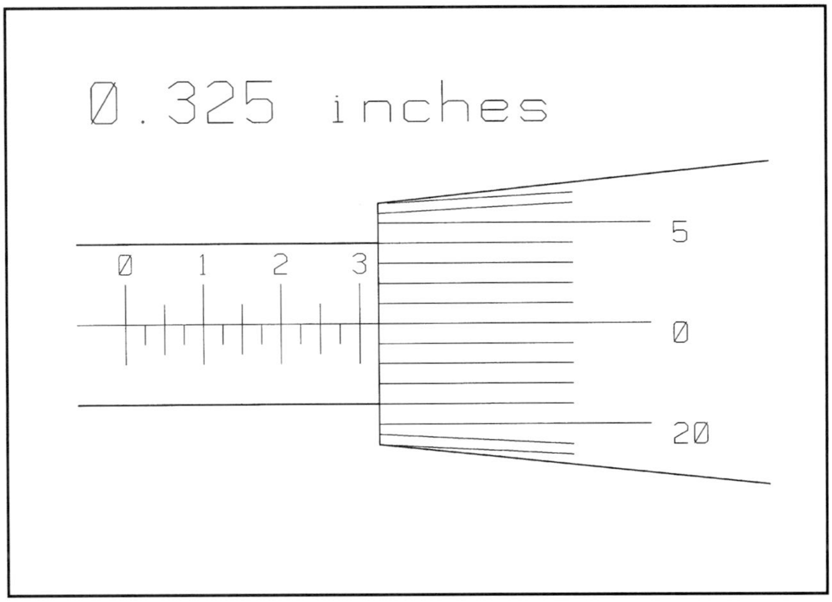

In this example, the 3 of the tenths-inch indicator is exposed, the barrel is right over the .025 indicator, and the barrel is aligned with 0. The measurement is 0.325in.

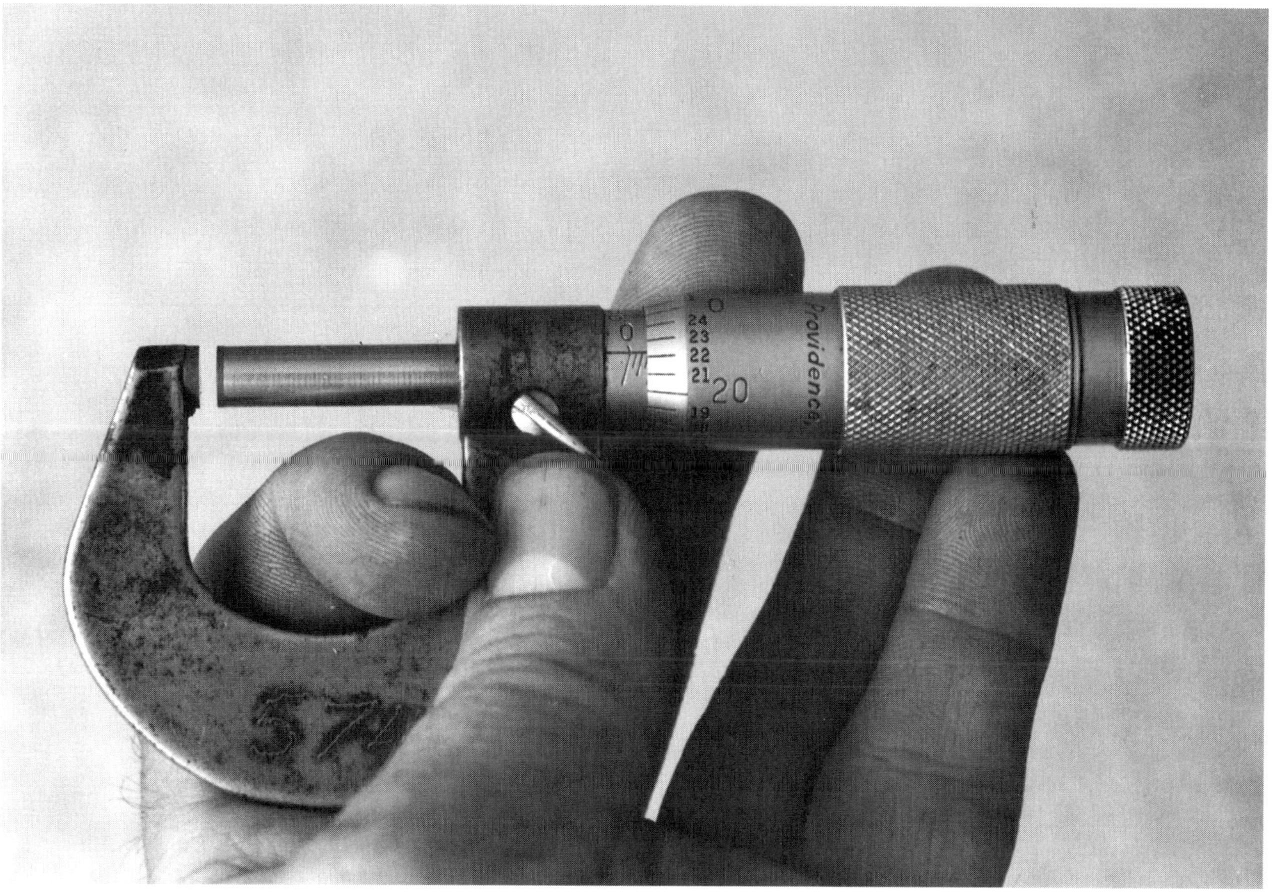

The reading on this micrometer is 0.0 + 0.05 + 0.022, or 0.072in.

ter to determine accurately the bore diameter.

Note: Using micrometers, inside micrometers, and split ball and telescoping gauges requires a little skill. The measuring instrument should be extended until it makes firm but gentle contact with surfaces being measured. Gently move the measuring device back and forth to ensure firm contact; that will give the proper measurement.

Dial indicator

The dial indicator consists of a precision gauge, usually marked at thousandths of an inch, which is moved by a plunger. When clamped or mounted firmly to the block or cylinder head, it can be used to measure the amount of front-to-rear movement or end-play in the camshaft or crankshaft. Not limited to this job, the dial indicator can be used to measure minute movement in virtually any component.

Additional jobs for the dial indicator include checking the flatness and run-out of rotating objects such as the flywheel.

Feeler gauges

Feeler gauges are thin strips of steel or brass machined to close tolerances. They are used

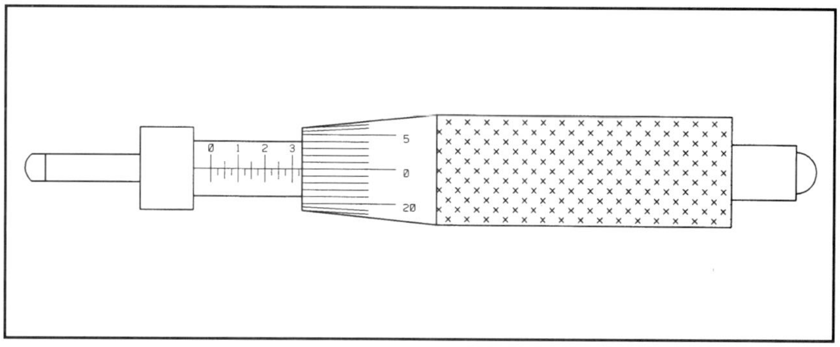

An inside micrometer can be used to measure the bore of the cylinder. Align the ends of the inside micrometer in the widest portion of the cylinder bore. The reading is the diameter of the bore.

to measure the gaps between components. Piston ring side clearance, crankshaft thrust clearance, and valve lash are among the gaps that can be measured with feeler gauges.

Plastigauge

A plastigauge is made of thin strips of plastic and is used to measure the rod and main bearing clearances. Remove the bearing cap and tear off a piece of plastigauge approximately the length of the bearing width. Place the strip of plastigauge on the bearing journal; install and torque the bearing cap. Remove the bearing cap and compare the compressed width of the plastigauge to the scale on the package. The compressed width indicates the oil clearance between the journal and the bearing.

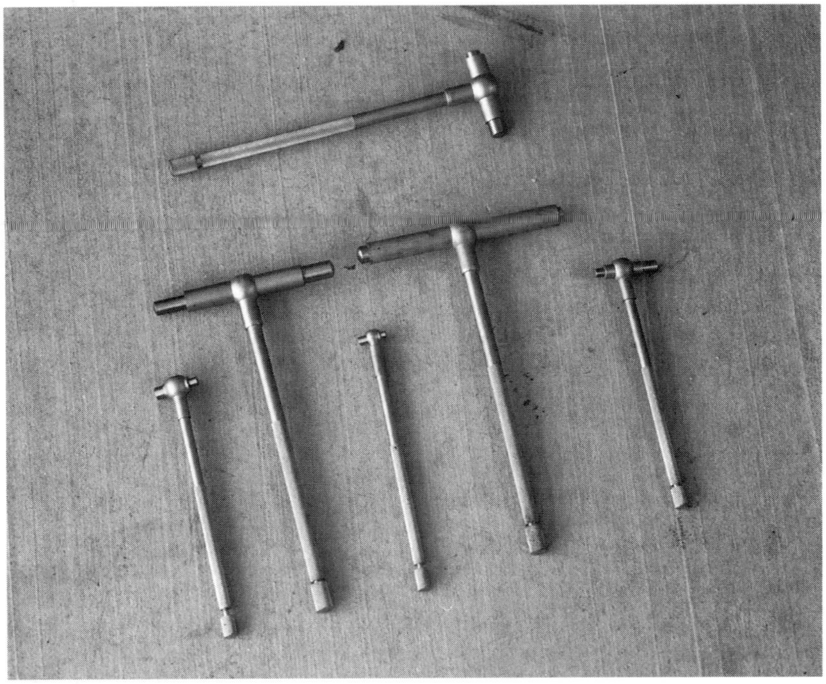

Telescoping gauges are a viable and economical alternative to inside micrometers. Align the ends of the telescoping gauge in the widest portion of the cylinder bore.

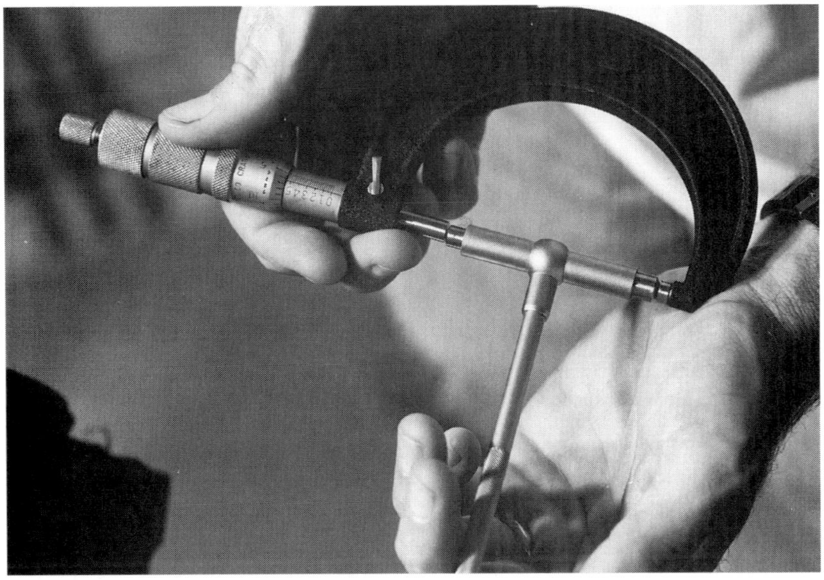

After the telescoping gauge is adjusted to the widest part of the cylinder bore, measure the length of the telescoping gauge with an appropriate micrometer. The length of the telescoping gauge will be the diameter of the cylinder.

Tachometer

The tachometer is used to measure engine rpm. While this tool is not essential for the actual rebuilding process, it is essential for adjusting the curb idle speed during the final tuning process.

Timing light

Like the tachometer, the timing light is a tune-up instrument, not an engine rebuild necessity. The timing light is used to synchronize the primary ignition system to the position of the crankshaft.

Cylinder hone

The cylinder hone is used to break the glaze that forms on the cylinder walls. If this glaze were left in place, the freshly installed piston rings would not seat properly. The result would be high oil consumption and reduced cylinder performance. The hone consists of abrasive stones mounted on a shaft. When mounted in a drill, the stones scrape the glaze off of the cylinder walls.

Proper use of the cylinder hone involves moving it up and down in the cylinder rapidly as the drill is running. The rapid up-and-down movement causes a cross-hatching on the cylinder walls. The cross-hatching accelerates the break-in of the new rings.

There are two common types of hones. The classic hone features three or four flat stones held against the walls of the cylinder by spring tension. Another type of hone is the "berry-ball" hone. Dozens of small abrasive stones are held against the cylinder walls by spring steel wires. Both hones are effective, but I recommend the berry-ball hone for the amateur or occasional rebuilder.

Timing wheel

The timing wheel is used to make precision adjustments in camshaft timing for performance. Cam timing is also critical to ensure that there can be a minimum amount of clearance between the piston and the intake valve. If the valve timing is not at its ideal, then piston-to-valve clearance may not be at its ideal either.

Ridge reamer

The ridge reamer is used to remove the ridge that builds at the top of the cylinder walls. This ridge forms during thousands of miles of operation as the rings move tiny bits of metal up the cylinder walls, depositing them at the top of the cylinder like the terminal moraine of a glacier.

CC gauge

The CC gauge is known to chemists as a graduated burette. A burette is a glass tube marked in milliliters. At the bottom of the tube there is a petcock. When the burette is filled with mineral oil and slowly drained through the petcock into the combustion chambers of the

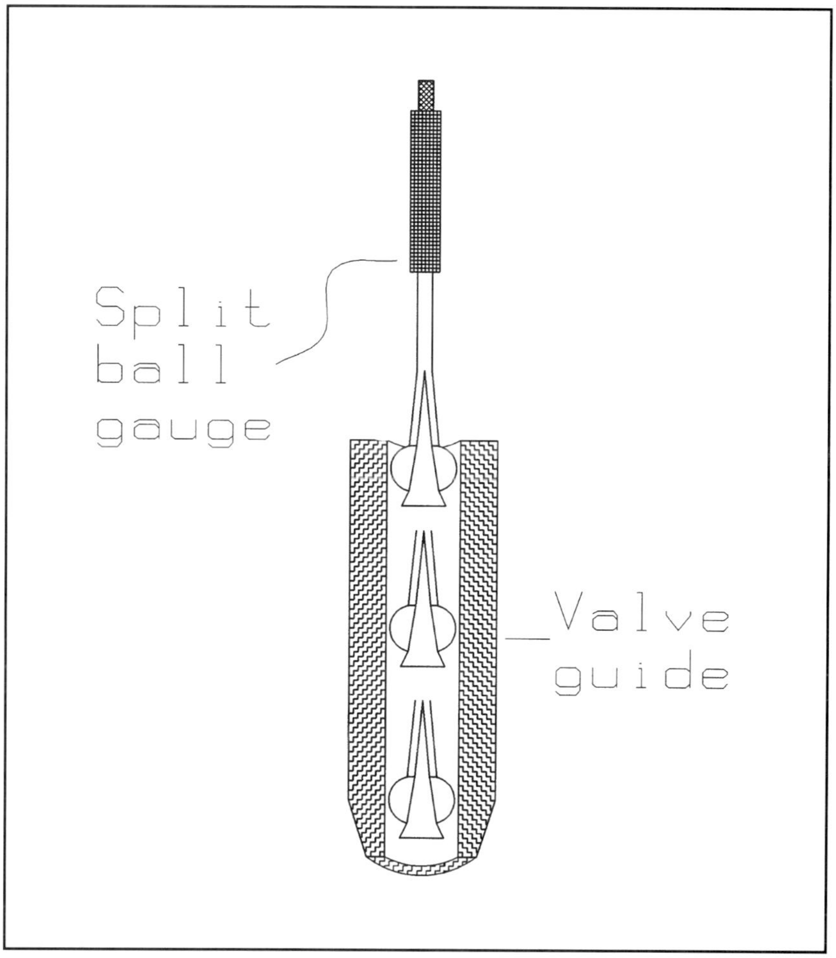

Split ball gauges are used like telescoping gauges but for the measurement of smaller critical bores.

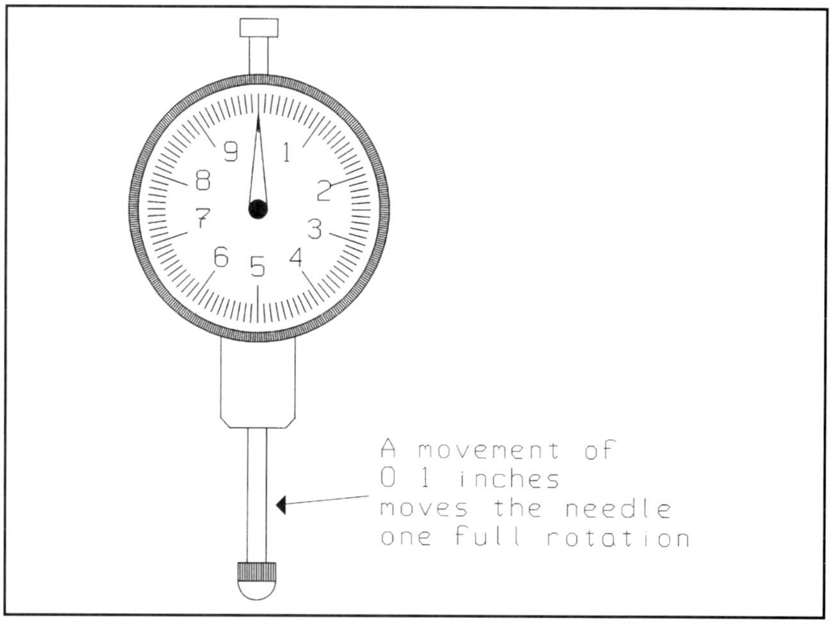

Right
Your dial indicator will get a healthy workout as you begin to rework the lower end of your engine. This measuring tool will do everything from helping to check the deck height of the block to helping phase the camshaft. A typical dial indicator measures in thousandths of an inch, and every rotation of the needle represents 0.1in.

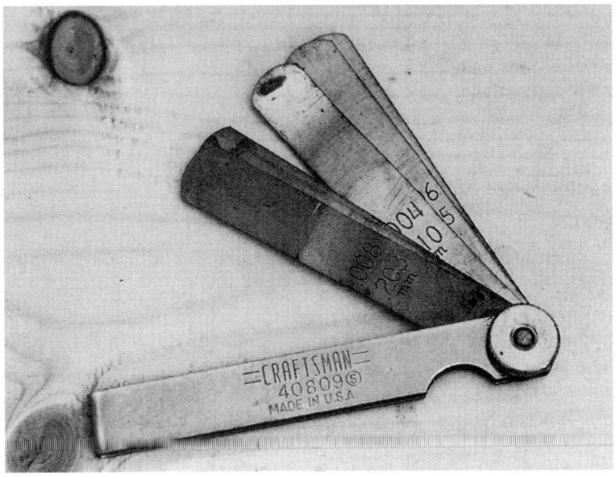

Feeler gauges are one of the most basic and essential automotive measuring tools. These feeler gauges have led a long and effective life. Since measuring with feeler gauges is done by dragging the gauges through tiny gaps, the gauges tend to wear and should be replaced regularly. These should have been replaced long ago.

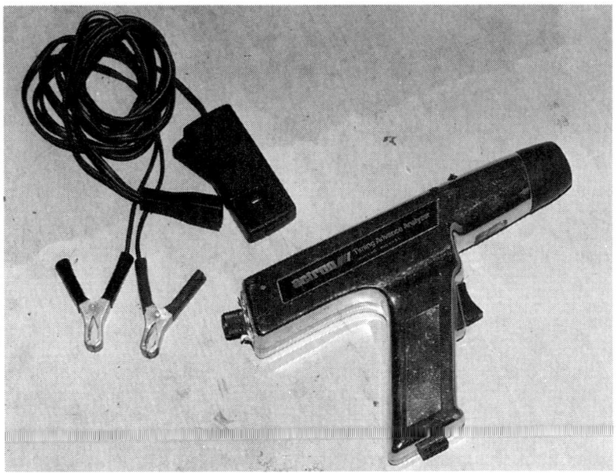

A tachometer and a timing light are essential for completing the performance makeover. This inexpensive timing light includes a timing advance feature. Many a quality overhaul has left the shop running "funky" simply because the technician did not take the time to properly adjust the timing.

cylinder head, it is possible to measure the volume of the combustion chambers. When blueprinting an engine, the equality of combustion chamber volume is critical to achieving equal power from each of the cylinders.

Common hand tools required

Make sure that you have a set of combination wrenches ranging in size from 1/4 to 1in; a 3/8in drive SAE and 1/2in drive SAE socket, ratchet, and extension set; and miscellaneous screwdrivers, pry bars, and hammers. If the engine is in a foreign car, you will need metric wrenches and sockets. If the car is a late model domestic, you will need to have both SAE and metric.

Machine shop tools

In addition to a collection of mechanic's shop tools, you will need to have access to a selection of machine shop tools. These tools are readily available in almost all communities, along with the skilled machinists to operate them. I live in a state with a population less than the city in which the reader probably lives. I live in a city with a population less than most neighborhoods in many American cities, and yet almost all of the machining services I could ask for are

Plastigauge is a handy measuring tool. It begins life as a mere sliver of plastic that is placed on the bearing journal surface of the crankshaft. Then the bearing cap is torqued to proper spec over it.

After the bearing cap is removed, the width of the plastigauge is compared to the marks of the protective paper jacket in which the plastigauge was purchased. This width indicates the bearing oil clearance. In this case, the oil clearance is 0.002in.

The cylinder hone is used to remove the glaze that forms on the cylinder walls during years of operation. There are basically two types of cylinder hones used by the technician. This is the industry standard which has been with us for decades, the flat stone hone. The second type is a "berry ball" hone which features dozens of little balls that scour the walls of the cylinder. A third type of hone is used by machinists to true the cylinder after it is bored.

right here in town (although I do have to be careful which shop I ask to do what). Before you get serious about engine building, you should scout out the following pieces of equipment and machinist skills.

Cylinder boring machine

If you are planning to use oversized pistons, or if the cylinders

Right
The timing wheel is used to set up the recommended phasing between the crankshaft and camshaft. Every cam grind has an ideal relationship between the opening of the intake valve for cylinder number one and the position of the piston in its stroke.

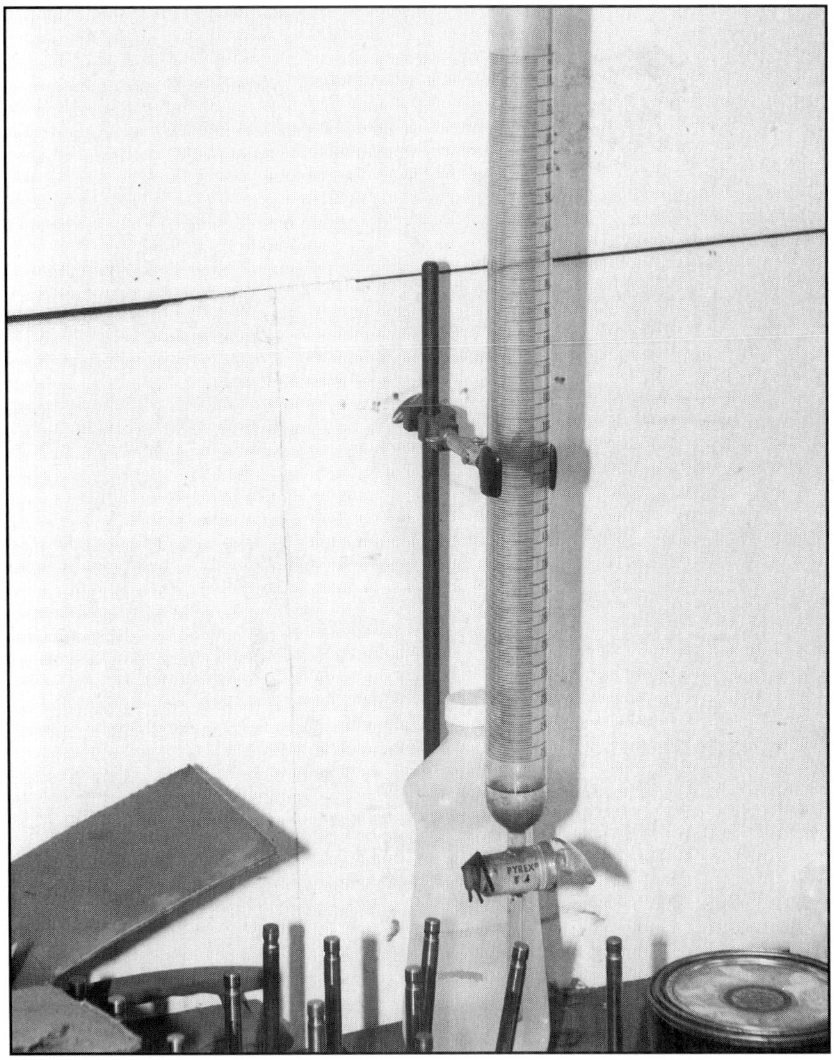

have been damaged by either prolonged wear or catastrophic failure, the cylinders may need to be rebored. This process enlarges and trues the cylinders. After this operation oversized pistons will need to be used.

Should the cylinder be severely damaged it may be necessary to sleeve the cylinder. This process involves greatly enlarging the damaged cylinder, inserting a steel sleeve, and boring the sleeve to match the rest of the cylinders.

Line boring machine

A seriously damaged block may require line boring. This operation is required when the main bearing bosses of the block are not in perfect alignment with one another. This problem is actually extremely common when the engine suffers a catastrophic failure and should always be checked during every overhaul when the engine is first disassembled.

To check bearing bore alignment, remove the engine block from the engine stand. Having the block in the stand during this measurement can distort the block and affect the accuracy of the readings. Place the machined straight-edge in the main

Above left
As the engine runs, metal and other particles are pushed to the top of the cylinder walls. When the pistons are removed from the engine with this ridge in place, there is a tendency to break rings and damage ring lands on the piston. The ridge reamer is used prior to disassembly to remove this ridge and eliminate the possibility of damaging the piston.

Left
A burette like this is essential when calculating the compression ratio of the modified engine. When filled with a fluid, that fluid can be slowly released into the combustion chamber of the cylinder head or onto the top of the head to measure their respective volumes.

bearing saddle of the block. Check for clearance between the straight edge and each of the bearing saddles. With a feeler gauge, the clearance should be less than 0.0015in (fifteen ten-thousandths).

Rod remanufacturing machine

When the engine is overhauled, each of the connecting rods should be checked to make sure that the big ends and small ends are aligned parallel on all planes. If they are not, they should be replaced. If replacing them is out of the question for either financial reasons or reasons of availability, you need to make sure your machine shop has the equipment to recondition the old rods. This is a fairly standard piece of equipment for serious automotive machine shops.

Crankshaft balancing machine

Even in the days when I was routinely rebuilding engines back to their stock new condition, I got in the habit of having the crankshaft balanced. It was a relatively inexpensive procedure, and I began to feel that it reduced customer complaints after the job was done. There are few things as disheartening as starting the freshly rebuilt engine, only to find it has a harmonic vibration. This is especially disheartening when you have spent many hundreds of dollars and hours making the rebuild as perfect in every other way possible.

Repairing a cracked block or cylinder

Ask your chosen machinist about repairing cracked blocks. Cracked blocks are common in some applications, climates, and uses. They are rare in others. If the block is cracked, there are several repair methods available to you, welding and threaded plug among them.

Welding the block is a highly specialized art; most machinists recommend replacement. If the block is a rare application, consult your machinist for recommendations on who is qualified locally to do the welding.

Sealing the crack with threaded plugs works well when the crack is small and replacement or welding of the block is out of the question. In this technique, the machinist will stop drill the crack, then drill a series of large holes along the length of the crack. Special plugs are then threaded into the holes. When threaded as far as they will go, they are cut off and ground flush with the block surface. Consult your machinist for advice on repairing cracked blocks.

This is a boring machine. What I mean is this is a machine used to bore out the cylinders of the block. Although it is a large and expensive piece of equipment, it is found in virtually every serious automotive machine shop.

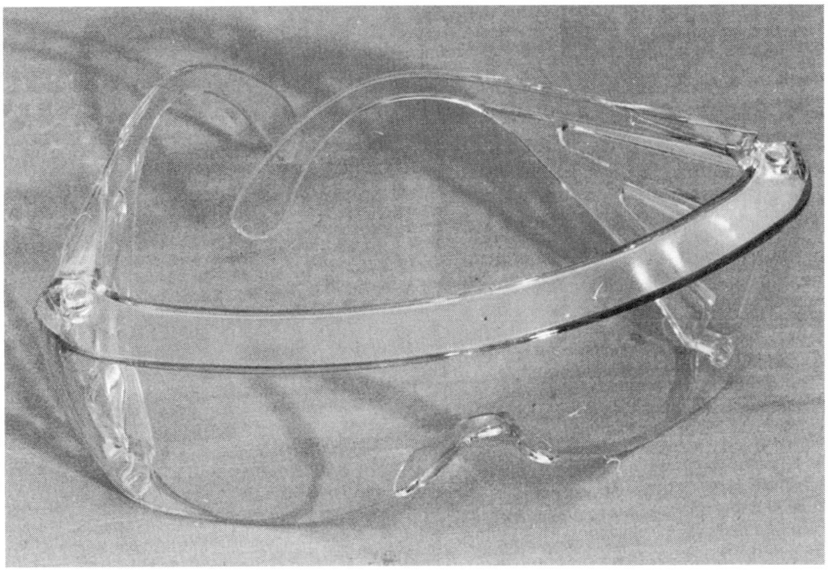

These may well be the most important tools you can purchase.

Chapter 2
Pistons

Compression ratio

Compressing air creates heat. This heat helps in ignition and raises the temperature of combustion. The compression ratio is the relationship between cylinder air volume when the piston is at bottom dead center and cylinder air volume when the piston is at top dead center.

Compression ratio is not nearly as cut and dried as I thought it was in my high school days. In the 1960s and early 1970s, compression ratios on stock engines straight from the factory were often as high as 10:1. Then came the fateful year of 1975. This was the first year for major control on exhaust emissions, especially oxides of nitrogen. Since oxides of nitrogen are created when combustion temperatures exceed 2,500 degrees F, and since high-compression ratios increase combustion temperatures, 1975 saw compression ratios suddenly drop to 7:1 or 7.5:1. Talk to those who traded their 1974 Dodge 318s for 1975 Dodge 318s. There was no comparison in the power of the two engines. The lower compression ratio of the 1975 model, along with a few other changes, robbed it of much power and fuel economy. Although the reduction in oxides of nitrogen was achieved, it was done at the expense of every definition of performance imaginable.

Calculating compression ratio

For those of us who struggle with math, there is fear associated with the word "calculate." For me, it brings back memories of Mrs. Maujer, my seventh grade math teacher at Jeanne d'Arc High School near Toul, France. Mrs. Maujer had the reputation of being merciless in her teaching of math skills. She was the one who taught me that one could have negative apples. Rest assured that calculating compression ratio is not nearly as complicated as what happens to Johnny's positive apples if Freddy combines them with his negative apples. First, we must know the total volume of the cylinder; this will include both the swept and unswept portions of the cylinder. The swept portion of the cylinder is that part where the piston travels; the unswept portion is the remaining volume of the cylinder and cylinder head that is never invaded by the piston.

Calculating the swept portion of the cylinder

Tradition has it that Euclid, a motorhead from the third or fourth-century BCE Alexandria, Egypt, was the first to calculate the swept volume of a cylinder. His theory was that if you square half the bore, multiply that times pi (P), then multiply the product of that calculation by the stroke, we would know the swept volume of the cylin-

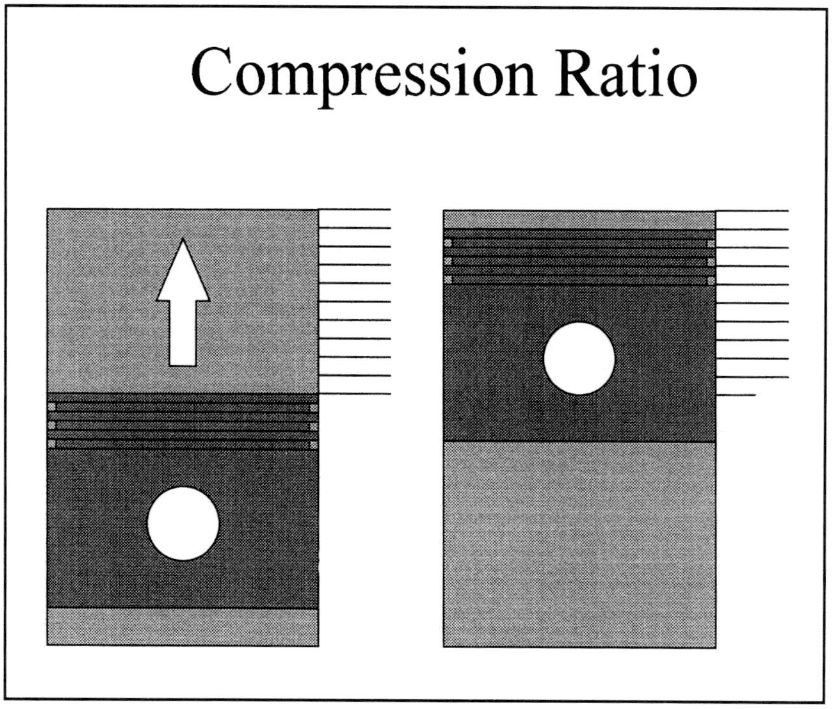

The compression ratio is the relationship of the volume of the cylinder measured with the piston at bottom dead center and the piston at top dead center. In the above example, the volume of the cylinder is eleven times greater with the piston at bottom dead center than it is when the piston is at top dead center. The compression ratio is therefore 11:1.

der. This was revolutionary at the time. Most of the competition was still moving the piston to bottom dead center, filling the bore with cream cheese up to the deck of the block, then scooping the cream cheese into the stomachs of goats. The typical big block of the day featured an average of three goat stomachs per cylinder.

Euclid's formula was as follows:

$(0.5 \times \text{bore})^2 \times \pi \times \text{stroke}$

Example:

$(0.5 \times 4)^2 \times 3.14 \times 3 = 4 \times 3.14 \times$
$= 12 \times 3.14$
$= 37.68\text{ci}$

In the example above, every cylinder has a swept volume of 37.68 cubic inches (ci). Therefore, each cylinder displaces 37.68ci of air as the piston moves from bottom dead center to top dead center. Multiply this result by 16.387 to convert cubic inches to cubic centimeters. The swept volume of this cylinder in cubic centimeters is 617.46. If this example were an eight-cylinder engine, it would be a 301.44ci (302) engine. (That is, a 0.052 cubic cubit engine.)

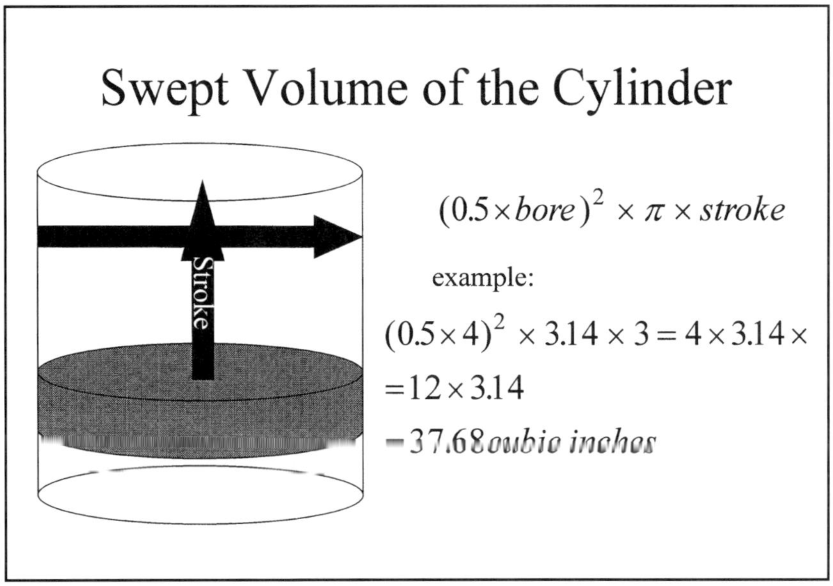

I first developed a fondness for the automotive performance and repair industry when my eleventh grade precalculus teacher told me that auto mechanics almost never have to use math. It was a trick. This business is a math-intensive business. The cylinder shown in the example above has a 4in bore and a 3in stroke. If there are eight cylinders, the engine is a 302ci engine. If there are only six cylinders, the engine is a 225ci engine.

Calculating combustion chamber volume

Now this is an exercise well suited for cream cheese. Instead of cream cheese, we use mineral spirits today. Find a piece of Plexiglas. Form a circle around the combustion chamber in the

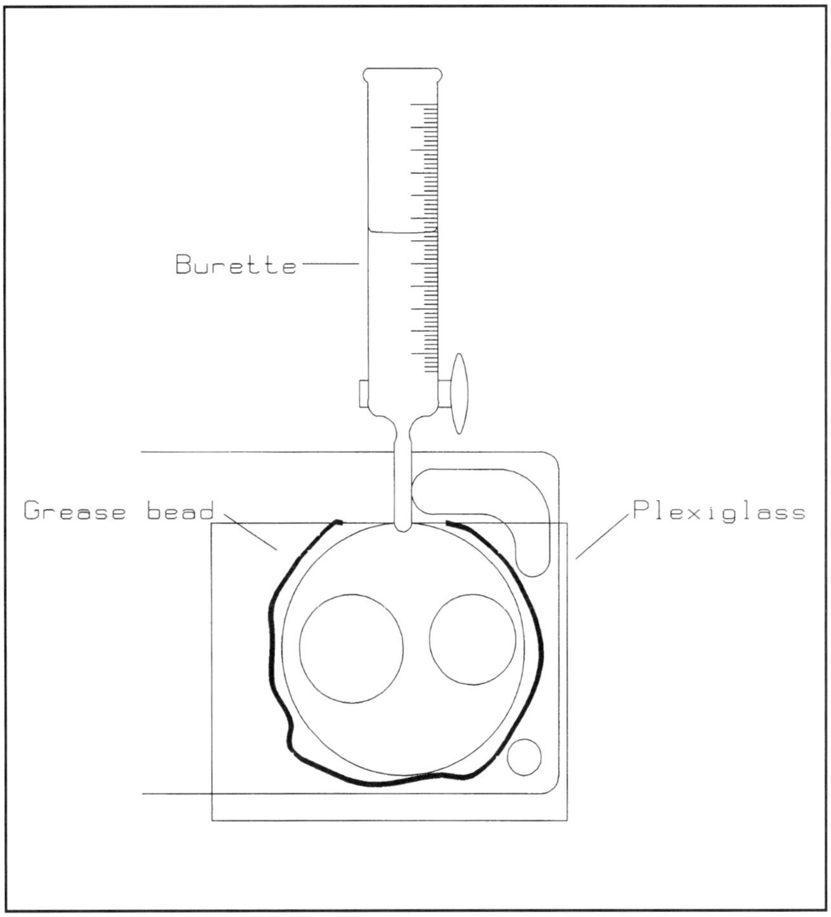

Right
In lieu of cream cheese, make a circle around the combustion chamber in the head with grease to form a seal when the Plexiglas is laid on the cylinder head. Place the Plexiglas on the head so that there is a small gap at the top of the mating point between the head and the Plexiglas. Using a burette, fill the combustion chamber with mineral spirits noting how much from the burette is used.

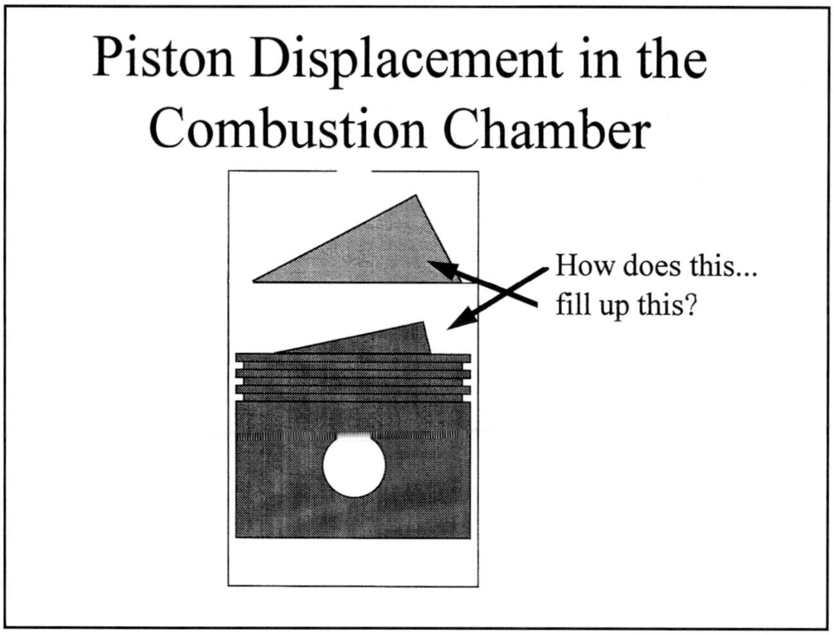

Since some pistons are not flush with the deck of the block when the piston is at top dead center, it is necessary when calculating the compression ratio to determine how much the shape of the head of the piston increases or decreases the volume of the combustion chamber.

head with grease to form a seal when the Plexiglas is laid on the cylinder head. Place the Plexiglas on the head so that there is a small gap at the top of the mating point between the head and the Plexiglas. Using a burette, fill the combustion chamber with mineral spirits noting how much from the burette is used. Let us say that the volume of our example is 96 cubic centimeters (cc).

Calculating piston displacement in the combustion chamber

Now if all pistons were flat on top, this step would not be necessary. Move the piston down the cylinder until it sits about 1/2in below the deck of the block. Make sure the highest point of the piston is below the deck of the block. Now use the burette to

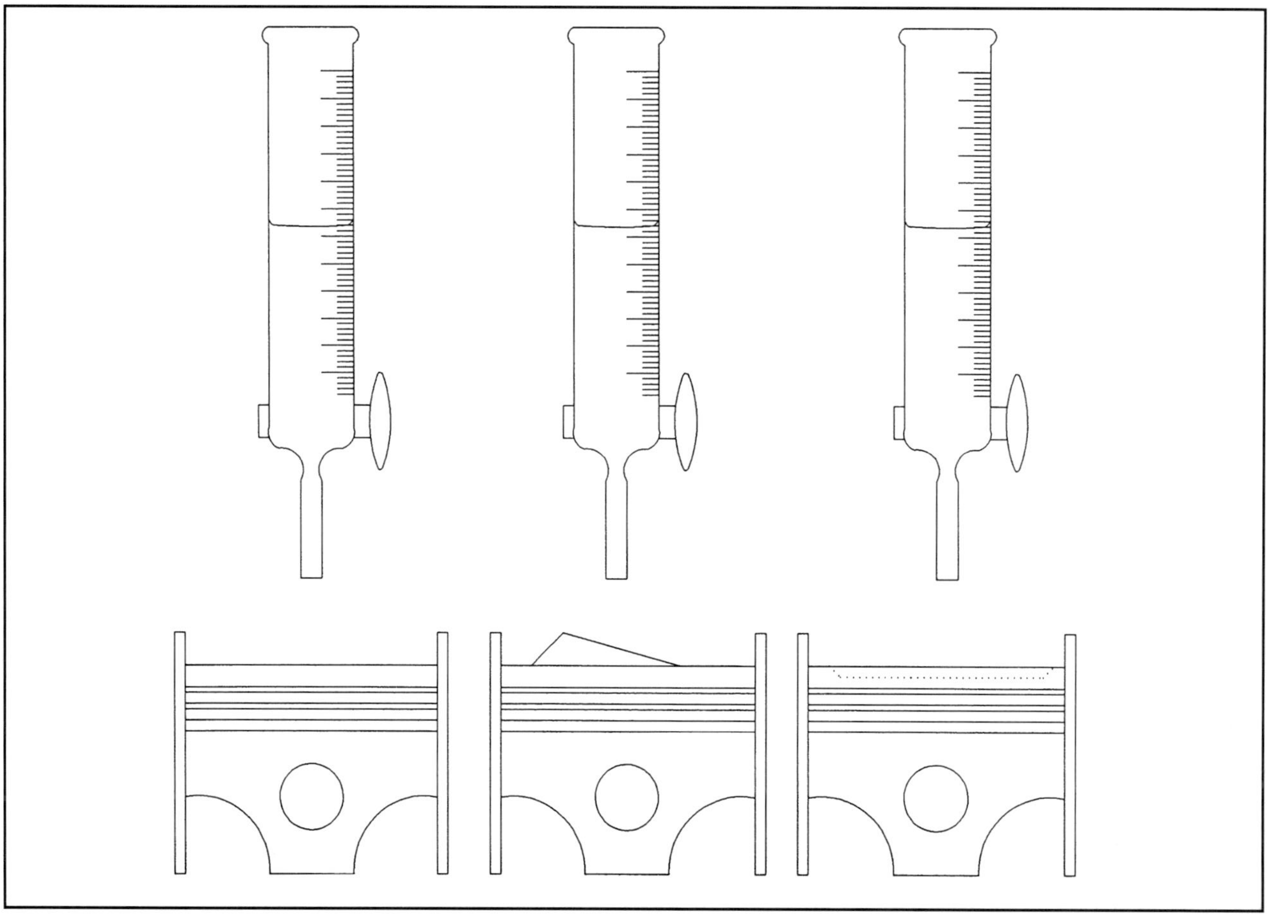

Final Calculation for Compression Ratio

$$\frac{\text{Swept volume} + \text{Combustion chamber volume} - \text{Piston displacement in combustion chamber}}{\text{Combustion chamber volume} - \text{Piston displacement in combustion chamber}} = \text{Compression ratio}$$

The final calculation of the compression ratio: add the swept volume of the cylinder plus the volume of the combustion chamber less the volume of the combustion chamber taken up by or added by the top of the piston. Take this total and divide by the combustion chamber volume less the volume of the combustion chamber taken up by or added by the top of the piston.

fill the cylinder. Record how much mineral spirits it took to fill the cylinder. Work fast because the mineral spirits will leak past the rings. If you prefer not to work fast or prefer to be as accurate as possible with your measurements, the leaking past the rings can be reduced. Carefully seal the gap between the piston and the cylinder wall with grease. As the grease is applied, be careful not to let any of it take up space in the combustion chamber. Now accurately measure the distance from the block deck to the top of the piston. Use Euclid's little formula to calculate how much volume the cylinder above the top of the piston would have taken up if the top of the piston had been flat. Now subtract the amount that flowed from the burette to fill the cylinder from the flat top piston volume. The difference is how much to subtract from the combustion chamber volume measured with the Plexiglas and burette. The difference is the actual size of the combustion chamber when the piston is at top dead center. For our discussion we will say that the top of the piston fills 10cc of the combustion chamber.

I suppose that it would be appropriate to note at this point that Mrs. Maujer was right; there is a place for math in the real world. If the piston top is depressed or recessed, then the piston design will add to the size of

Left
Use Euclid's little formula to calculate how much volume the cylinder above the top of the piston would have taken up if the top of the piston had been flat. Now subtract the amount that flowed from the burette to fill the cylinder from the flat top piston volume. The difference is how much to subtract from combustion chamber volume measured with the Plexiglas and burette. The difference is the actual size of the combustion chamber when the piston is at top dead center.

the combustion chamber. It might be said the combustion chamber size is reduced by a negative amount.

The final calculation for compression ratio

The final calculation for compression ratio is the swept volume plus the combustion chamber size; the sum is then divided by the combustion chamber size. In our example, the swept volume was 617.46cc. The combustion chamber size was 96 less the 10cc taken up by the piston. The total size of the cylinder, swept volume plus combustion chamber, is 617.46 plus 86 or 703.46cc. Divide the 703.46 by 86 and you get a compression ratio of 8.18 to 1.

Valve relief volume

The valve relief in the pistons lowers the compression ratio. The volume of these areas needs to be added to the volume of the combustion chamber in the compression ratio calculation. The volume of the reliefs is easy to measure; the technique is similar to the technique used to measure the combustion chamber volume itself. Fill the valve reliefs with the burette. Monitor the amount of oil, or for that matter, cream cheese, that it took to fill the reliefs. Add that volume to the volume of the combustion chamber before doing your final compression ratio calculation.

Power and compression ratio

I began this discussion of compression ratio by talking about my adolescent misconception of its importance. There is little doubt that a confined burn will make a bigger bang than an unconfined burn. The real question is, does a bigger bang mean more power? As far as big bangs go, a former racing foe of mine, Alf Gephardt, had a BMW 2002 that took the cake. The unofficial ground rule of the Green Valley Racing Association was to never let Alf pass on your left. Over the sound of your own engine, over the noise suppression of your Bell Star helmet, over the cotton stuffed in your ears, Alf's little Beemer was still deafening. Its bark was much worse than its bite, and its bite was considerable.

Compression ratio has to do with how confined the charge is when it is ignited. The larger the confined space, the less pressure will be created when ignition occurs. The smaller the confined space, the greater the pressure that will be created when ignition occurs. Now imagine a roomful of undisciplined children. The passion with which you desire to leave the room is inversely proportional to the size of the room. You want to get out; the smaller the room is, the more you want to get out. The same is true of the burning gases inside the combustion chamber. The more passionate the burning gases are, the greater the pressure they create and the more force is generated.

To give you a more graphic representation of what this means, imagine a 300-square-foot (sq-ft) room. This will represent the volume of the combustion chamber when the piston is at bottom dead center. Now put thirty kids in the room with you. Begin to move the walls closer together as you feed the kids jelly doughnuts and Kool-Aid. By the time the moving walls have shrunk the room to 30sq-ft, representing the piston at top dead center, each of the kids has consumed three doughnuts and 16oz of Kool-Aid. There is pressure in that room, wouldn't you say?

The bad part about high compression

For most things in life, there is both a good side and a bad side. This is also true of compression. High-compression engines survive well on high-octane fuels. These fuels, such as

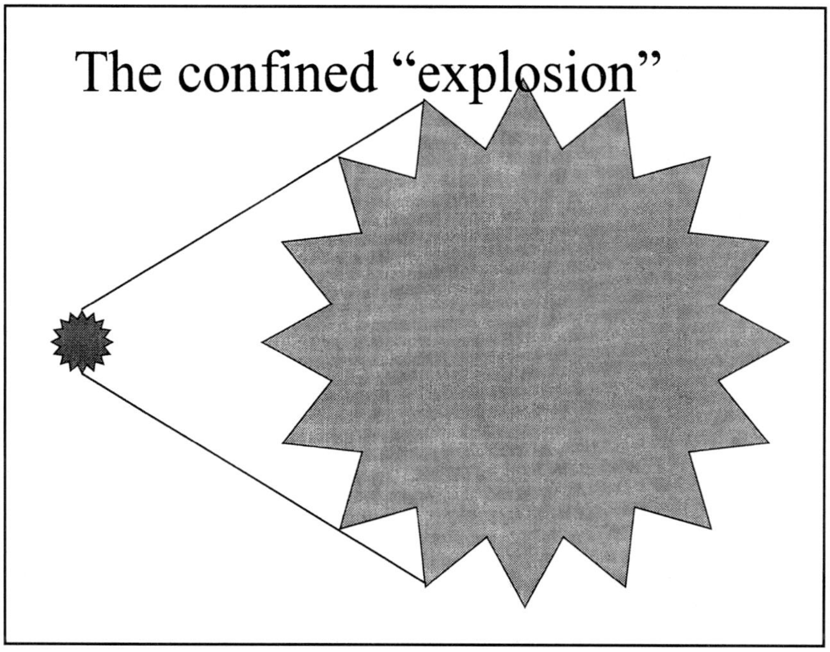

When the fuel is ignited in the combustion chamber, in combination with the air, it begins to expand. The more confined the "explosion" when it begins, the more relative expansion can occur. The more relative expansion, the greater potential there is for power.

methanol, ethanol, natural gas, and the mythical high-octane gasolines of the 1960s burn, or burned, slowly and evenly in the combustion chamber. The slow even burn reduces hot spots and produces a more even temperature in the combustion chamber. A more even temperature in the combustion chamber yields a more even expansion of gases and more even temperatures. For high-compression engines this is particularly important, since the average temperature of combustion is already elevated. With an elevated average temperature, the spot temperatures are close to or past the point that melts metal. Today's gasoline is not very compatible with high-compression engines. High compression and low octane can be death to an engine.

Several years ago, while working as an import car technician at an independent shop, a six-cylinder Mercedes was brought into the shop by the owner. The primary complaint was a lack of power. The compression test indicated that four of the six cylinders had no compression at all. When the cylinder head was removed, the pistons in these four cylinders had holes in them at least the size of quarters. These holes had been burned into the pistons by high temperatures associated with detonation. The most likely cause of the detonation was a relatively high-compression engine running for an extended time on low-octane fuel.

Detonation occurs when the air/fuel mixture in the combustion chamber ignites prematurely. This early ignition may be caused by over-advanced timing, high-compression, low flash point for the fuel, or sharp, glowing edges in the combustion chamber. When the air/fuel mixture ignites with the piston on the upward stroke, the gases resulting from combustion begin to expand rapidly. The piston, therefore, must compress combustion gases rather than simply the relatively low pressure of air that has been drawn in from the atmosphere. One of the results is the classic "marbles in a tin can" sound familiar to all who have had the pleasure of detonation. Another result is that the rapidly expanding heated gases are further pressurized, and therefore heated, as a result of their confinement. The final result is extremely high temperatures and pressures, which can each damage pistons and other expensive components.

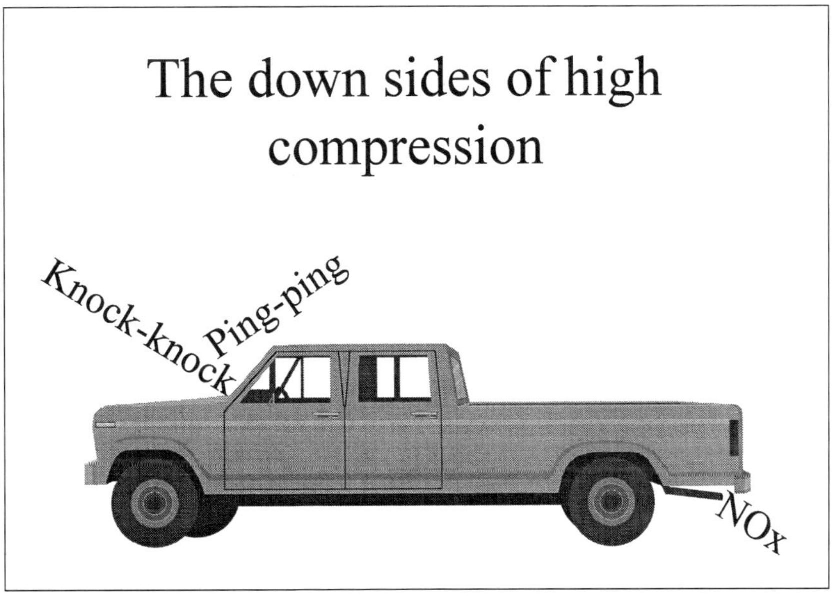

High compression in an engine is not all good. High compression causes high-combustion temperatures. High-combustion temperatures can result in engine damaging pre-ignition. Additionally, the output of oxides of nitrogen will tend to increase.

Detonation is a clue that something more sinister is happening. The same high pressures and temperatures that produce detonation also cause oxygen to combine with nitrogen, resulting in the production of NOx. This was, as mentioned earlier, the main reason compression ratios were lowered in the 1970s. In the late 1980s and early 1990s, computer-controlled timing systems were developed to reduce the possibility of detonation. This also reduced the instances in which the conditions were right for the formation of oxides of nitrogen. This development allowed the compression ratios to creep back up during the mid-1990s.

Combustion chamber design

There are many designs in combustion chambers. Some are designed to concentrate energy, some to reduce emissions, and still others for the simple convenience of manufacturing. Here are a few details about just a few of the head designs. For more information about cylinder heads, consult my book *How to Build and Modify Cylinder Heads, Camshafts and Valvetrains* or *Chevrolet Small-Block V-8 Cylinder Heads* by David Vizard.

Wedge

This is not the oldest cylinder head design. It is the cylinder head design that got the valves in the cylinder head rather than in the block. The old "flat head" design limited the compression ratio and reduced the efficiency

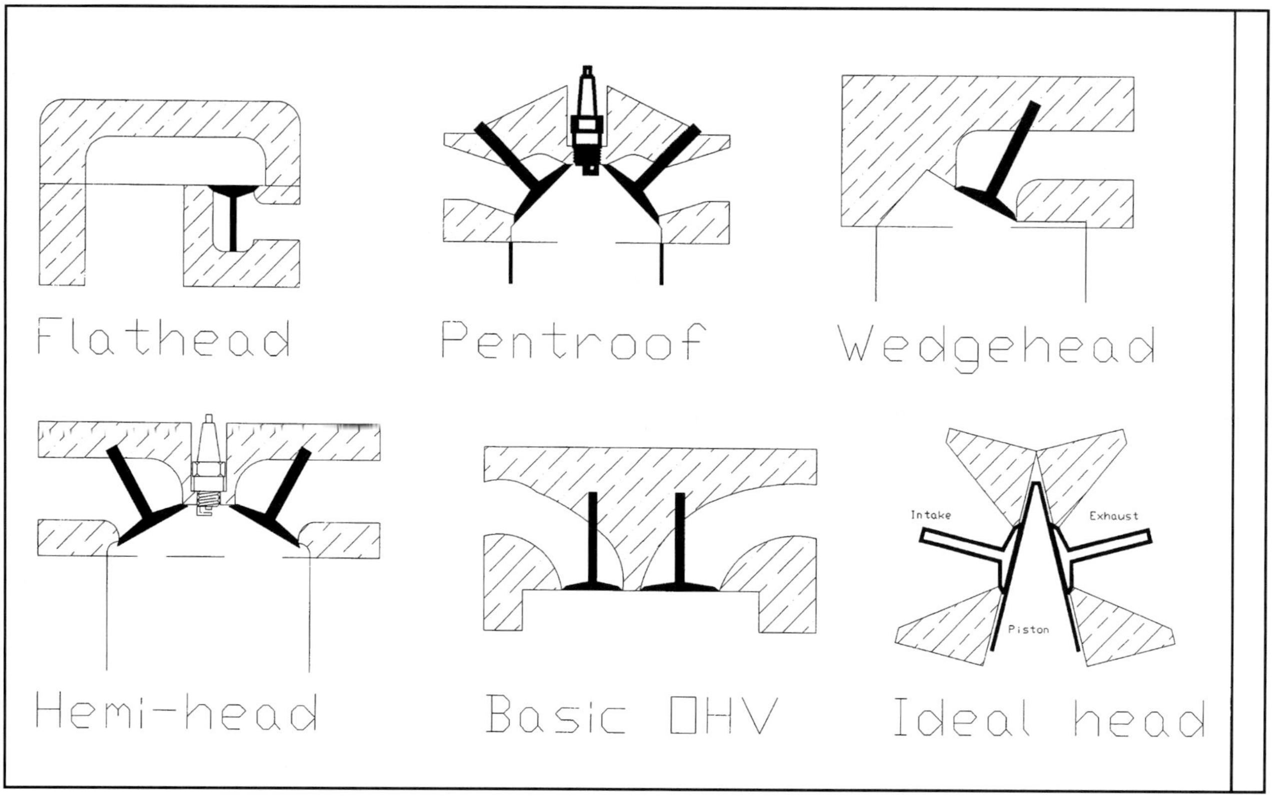

There have been many popular designs in cylinder heads and combustion chambers. All of these have advantages with respect to one another, but none is the ideal head. Notice that in the ideal head, cross-flow is maximized and the shape of the piston allows for maximum compression ratio. The so-called ideal head has its own faults, however. The valves themselves mask air and exhaust gas flow, for instance.

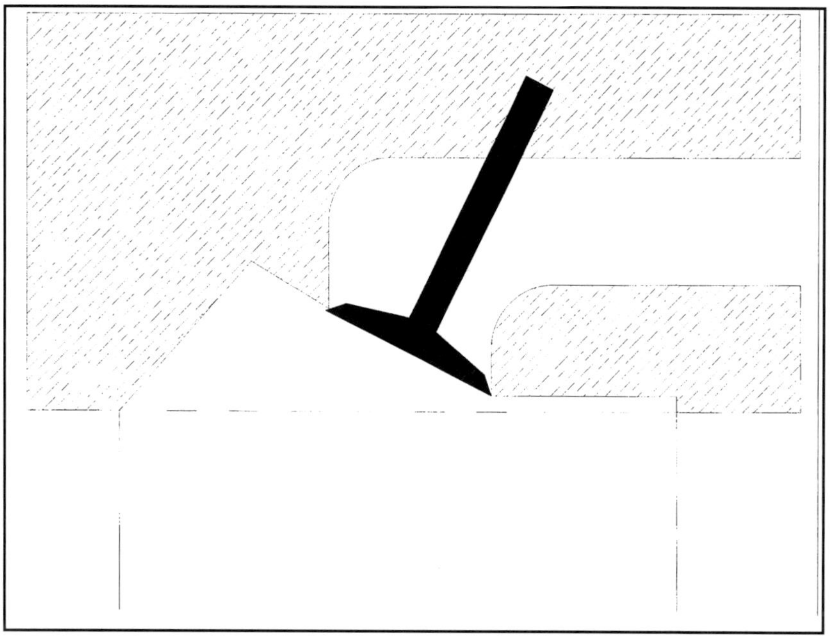

One of the most common cylinder head designs for the past several decades is the "wedge." Mechanically efficient, the flow of the gases through the cylinder head and combustion chamber requires a u-turn between intake and exhaust.

of air flow. The wedge head design dates to the 1950s, when the manufacturers first began to turn the family sedan into a land rocket. These combustion chambers are the classic overhead valve combustion chambers.

Hemispherical

They say that anyone who remembers the sixties misses them. That may or may not be true; however, anyone who remembers the sixties and was into performance cars remembers the Chrysler hemis. The unique, for its time, cross-flow design of the combustion chamber allowed for phenomenal intake velocity and exhaust gas scavenging. Shortly after graduating from high school, one of our car-head gang lowered himself to buy a Toyota Corolla with a 1600cc engine. Randy was immediately black-listed, left out of the gang activities. Then we learned the 1600cc Toyota engine had hemispherical combustion chambers. Randy was immediately repatriated. In fact, he was the only guy in the group to have a hemi! Since this time, the hemispherical combustion chamber has been used to both increase engine power and increase fuel economy.

Stratified charge

The best known of the engines with stratified charge combustion chambers is the Honda CVCC engine. CVCC means Centrifugal Vortex Combustion Chamber. The spark plug ignites a very rich charge in a precombustion chamber. This mixture burns slowly and provides a heat source for the continued ignition of a lean (stratified) mixture in the main combustion chamber. The biggest advantage of this design is low emissions.

Pent roof

This is one of the latest combustion chamber designs. The pent roof design features a steep angle between the intake and exhaust valve seats. Used on many multivalve engines, the design provides for excellent cross-flow of intake and exhaust gasses.

Heat dissipation

Of all the engine components, the one that has the roughest life is the piston. The piston must transfer the power created by the explosion in the combustion chamber to the connecting

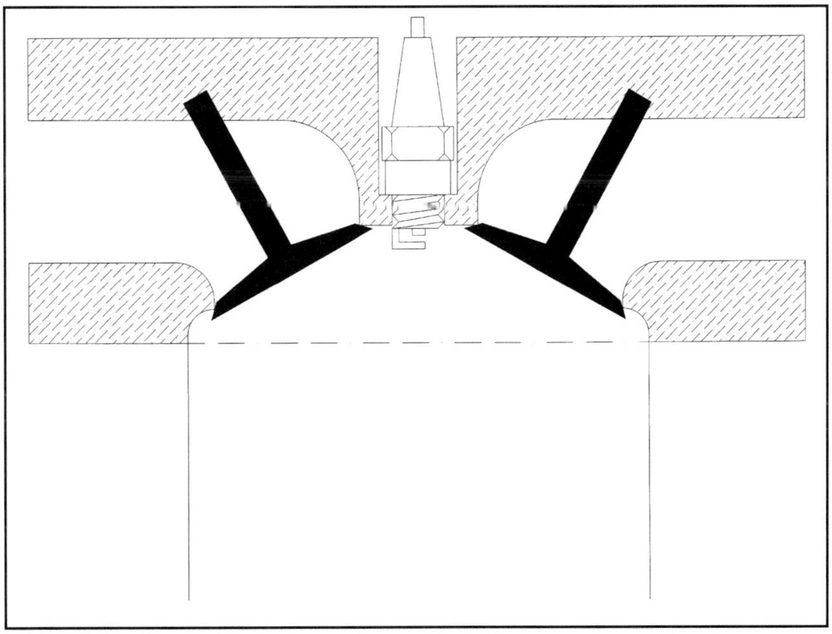

The "hemi" was the mythical engine of my youth. The valves in this head arrangement are situated on opposite sides of a dome. This provides good cross-flow, thereby optimizing intake air flow and exhaust gas flow.

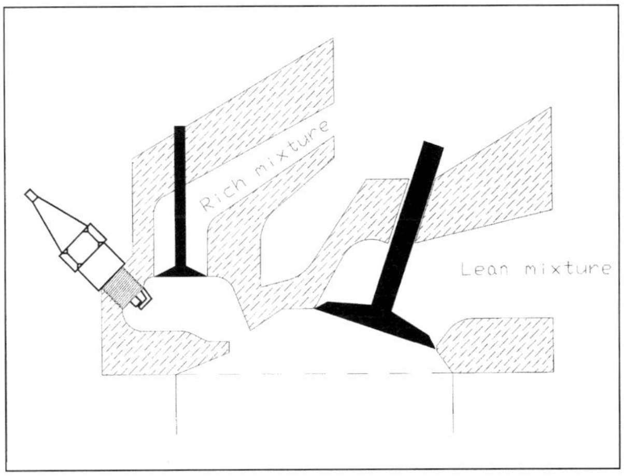

Stratified charge combustion chambers were popular during the 1970s. These cylinder heads featured two intake paths to the combustion chamber. One path was fed with an extremely rich mixture designed to sustain combustion. The other carries the primary air fuel charge; this is the primary source of gases to be expanded after ignition. The main advantage of this design is a cool combustion temperature and low emissions.

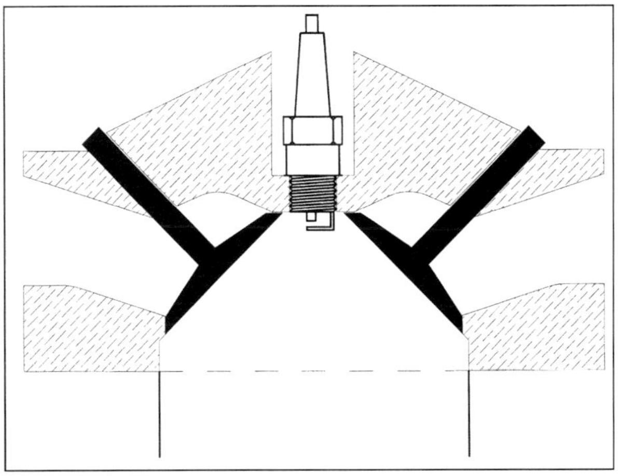

This is one of the latest combustion chamber designs. The pent roof design features a steep angle between the intake and exhaust valve seats. Used on many multivalve engines, the design provides for excellent cross-flow of intake and exhaust gases.

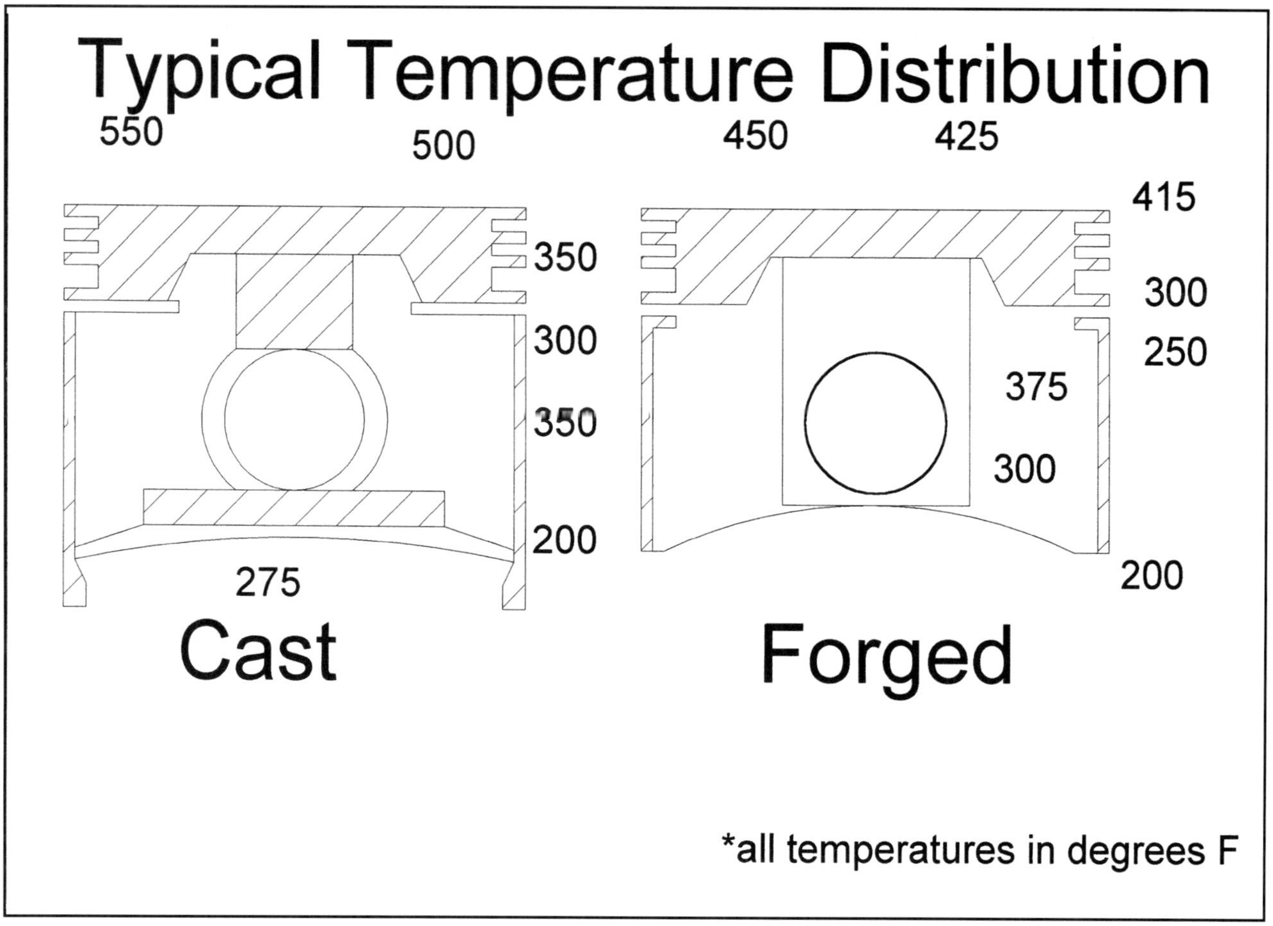

Pistons must be able to withstand extremes of temperature. At times the piston will be sitting stationary in the cylinder at a temperature of -40 degrees F or less. Suddenly the piston moves up the cylinder and there is an explosion exposing the piston to temperatures hundreds of degrees higher. Even within the piston itself temperatures can vary by over 200 degrees, even after the engine has reached operating temperature.

rod, which then must transfer the power to the crankshaft. The first item to be assaulted in this chain of power transfer is the piston. When you really think about it, the life lead by a piston is uniquely difficult. Picture yourself as the piston in a 1994 Ford Mustang 302. Your owner has just been transferred to Fairbanks, Alaska. On the morning of January 27 you are awakened by a rude electric starter motor. The intake valve opens as you begin your descent in the cylinder. Air blasts in at 40 degrees below zero. Soon you find yourself being shoved upward; the air on top of you is rapidly rising in temperature. Suddenly the spark plug fires and the gases rise in temperature by more than a thousand degrees. Your topside is heated to nearly 600 degrees. This high temperature lasts only a moment; soon the gases begin to cool and are quite cool when you shove them out of the exhaust valve. In another moment, the intake valve opens to admit another blast of 40 degree below zero air. This cycle continues until the engine is shut off. I find it amazing that pistons do not disintegrate all the time.

Since the piston has varying thicknesses, the heat will be transferred at varying rates. The temperatures within the piston can vary by 300 to 400 degrees and even more at any given time. Many engine components would simply self-destruct.

Thick ridges are built into the piston. These ridges are used to carry heat from the head of the piston to the skirt where it can more easily dissipate. Additionally, most pistons are not round when they are cold. The piston is designed to expand in such a way that it becomes round at operating temperature.

Piston types
Materials

The most common material for pistons is aluminum. Many older pistons were made of gray cast iron. The ideal materials will have a high melting temperature and be light. One of the

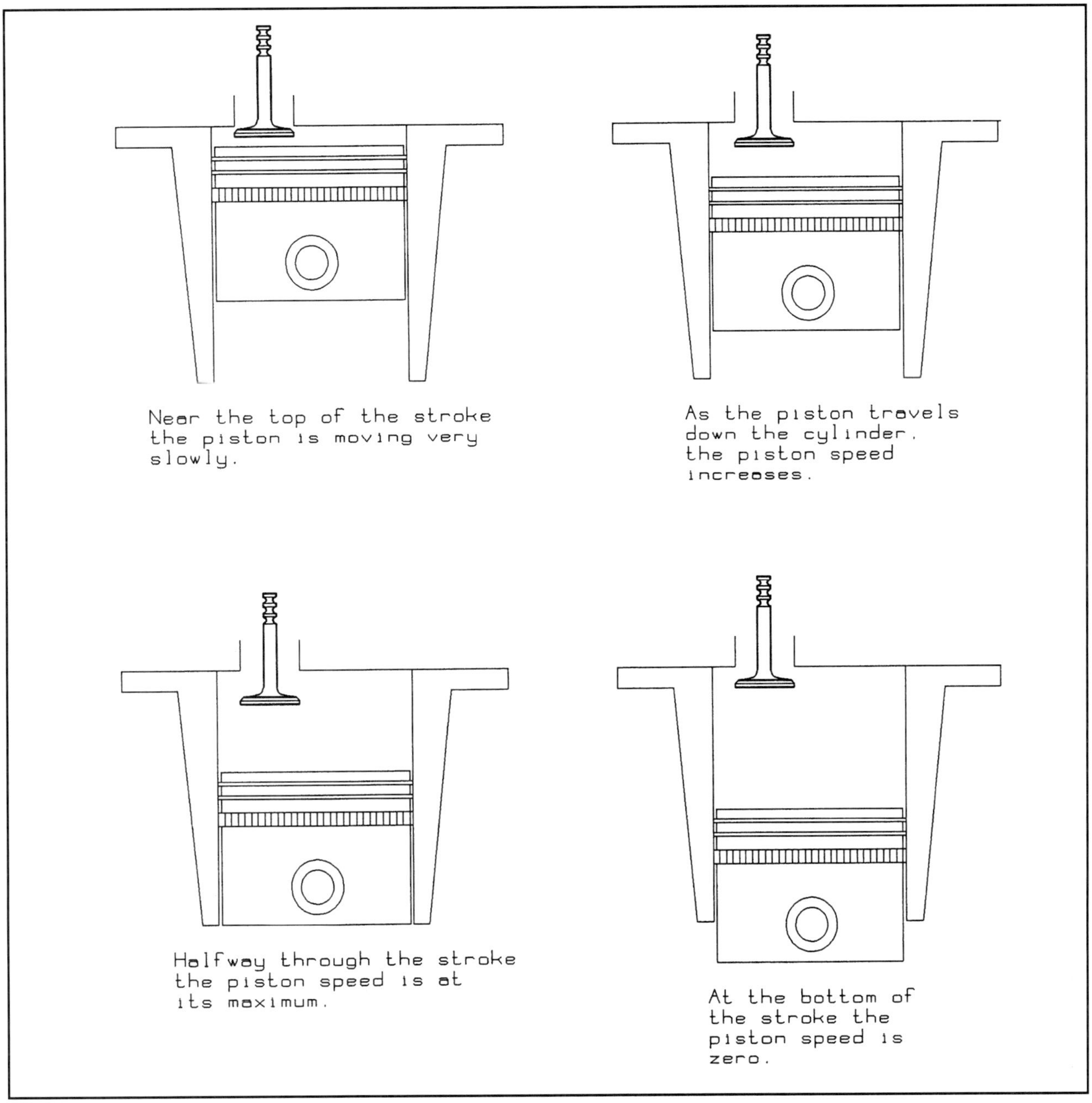

Pistons are incredible components. At bottom dead center they are sitting at 0mph. As the piston rises toward the top of the stroke, it accelerates dramatically. By the time it has traveled halfway up the cylinder, it can be traveling as fast as 45mph. If this rate of acceleration were continued, it would take less than 15.54 hours to achieve light speed. Incredible.

primary factors that limits the maximum speed of the engine is the mass versus the strength of the pistons.

Several years ago, I heard a story from a mechanic responsible for maintaining heavy equipment in the Canadian bush. The operator of a bulldozer had called in a complaint about a loud knocking noise from the engine of his machine. The mechanic packed up his toolbox, hopped a flight in the company Cessna, and arrived at the site of the bulldozer several hours later. He and the operator started the engine. It hummed like a new bulldozer. They operated the bulldozer for several hours, after which the mechanic pronounced judgment that the noise had been "a piece of carbon bouncing around in the cylinder."

The mechanic called for the Cessna to return and retrieve him. Several hours later he returned to the home base shop only to be faced with the message that the problem had returned only a few minutes after he left. Returning to the bulldozer site, the mechanic decided to just tear the engine down. As soon as the cylinder head was removed, there was an obvious problem—a six-cylinder engine with three pistons near or at top dead center. Normally on any six-cylinder in-line engine there should be only two pistons at top dead center. This was instantly interpreted by the mechanic as a problem. On further investigation, the mechanic found that the top of one of the pistons was left behind at the top of the cylinder, sometimes, having completely separated itself from the rest of the pistons. When the engine was shut off, if the broken piston stopped at near top dead center, then when the engine was started the piston would not create pressure; there would not be enough compression heat to cause ignition, and the piston top would remain at top dead center.

While this was an extreme example, and a diesel at that, it does illustrate how the piston is expected to accelerate from 0 to 30mph then back to 0 in a distance of 3 or so inches as much as 160 to 180 or even more times per second. Remember that on every crankshaft revolution, the piston accelerates and decelerates twice, once from BDC to TDC and once from TDC to BDC. As the speed of the engine increases, so does the desire of parts of the piston to remain at the top, or bottom, of the cylinder. A piston with a head weighing only 4oz subjects the thin connection between the head and skirt to over 300psi of force as the piston accelerates toward the midpoint in the stroke.

We ask the pistons to make 400 million strokes and to journey about 19,000 miles (3 to 4in

All of the abuse we heap on the pistons of the engine, and then we complain when there is a problem. This engine had been in service since 1962, powered the car day after day, through the Cuban Missile Crisis, through the Kennedy assassination, through the Gulf of Tankin, through Woodstock, the breakup of the Beatles, Vietnam, disco music, Reaganomics, and Rush Limbaugh. How many readers can honestly say the same?

at a time) in 100,000 miles of engine operation, and yet we complain when the engine wears out. Extreme operating conditions are the norm for pistons, even in a normal street engine.

The pistons in a high-performance engine may be asked to perform and hold together under even more adverse conditions. Never mind that the piston must accelerate from 0 to 45mph and decelerate from 45 to 0 up to 16,000 times per minute, 266 times per second (at 8000rpm), but it has to stay together as it does so at temperatures exceeding 2,500 degrees F. Racing pistons are truly an amazing feat of metallurgy. Many racing pistons will have a great deal of their total mass, the piston head, attached by very thin and flimsy ring groove areas. These areas must be able to transfer the force of the piston being driven down by the expanding gases of combustion during the power stroke. They must also be able to overcome the inertia of the head of the piston resting at top dead center at the beginning of the intake stroke and drag the head of the piston down the cylinder.

Metallurgy of pistons

Because the pistons are subjected to such severe and extreme conditions, great attention is given to the metallurgy of the pistons. The aluminum alloy used in most pistons contains silicone, copper, and manganese, in addition to aluminum. Variations in these blends are often what make a particular brand of piston unique.

Forged or cast?

Cast pistons are the piston of choice for most uses. The main cause for their popularity is their relatively low cost. If the engine is being built to regularly exceed 5000rpm, then a forged piston would be more desirable. Forged pistons have an extremely dense grain structure and can be up to 70 percent stronger than cast pistons. Forged pistons have the disadvantage of being more expensive than the cast piston. Of course, you have probably figured that out for yourself by now. As most things go, the faster or better the parts, the

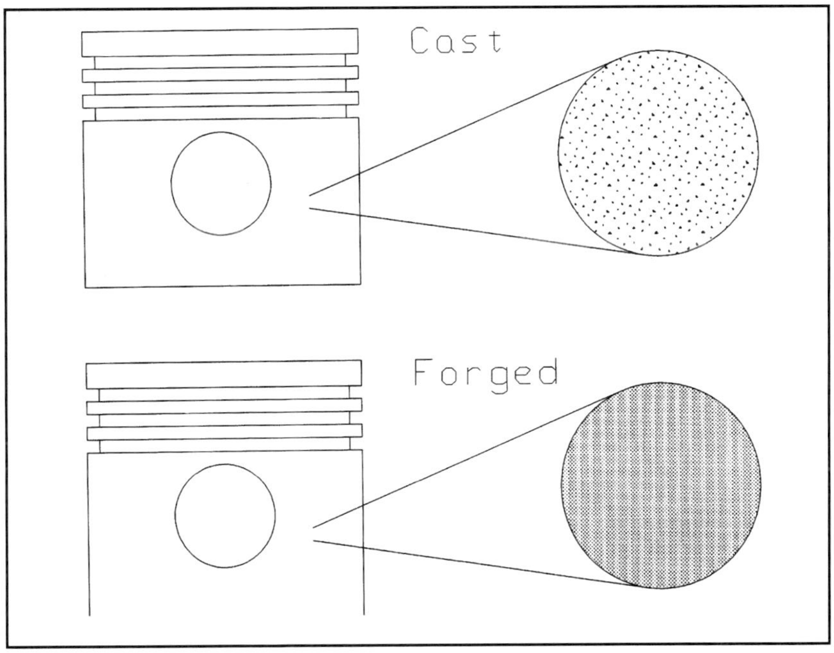

Cast pistons have a grainy, somewhat disorganized structure. Forged pistons have a denser and more organized grain and structure. The forged pistons, therefore, are the stronger structure.

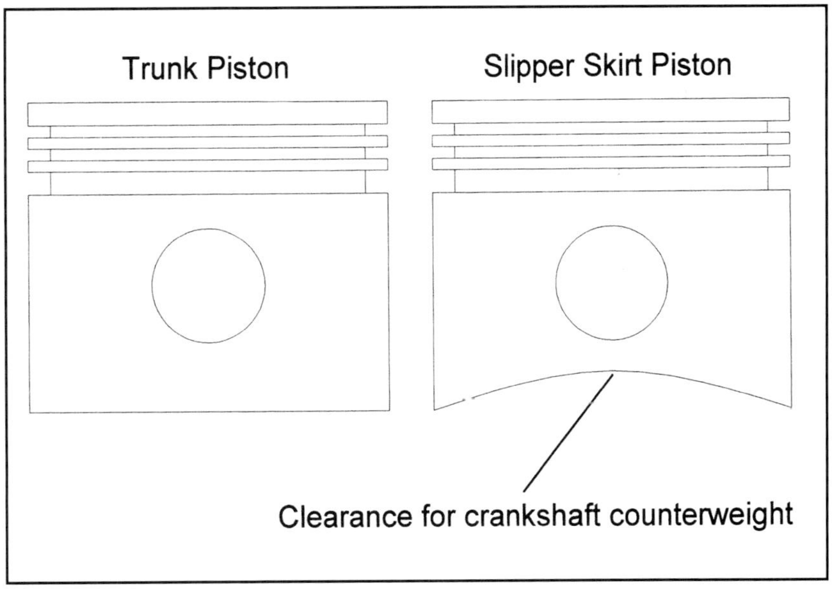

There are two primary skirt designs for pistons. Trunk, or full-skirt, pistons date back to the time of Henry Ford and even earlier. When counterweights were added to balance crankshafts that did not inherently balance themselves, clearance had to be provided at the end of the piston skirt for these counterweights. Enter . . . the slipper-skirt piston.

Above is an example of both a slipper-skirt (left) and a trunk-skirt (right) piston.

more expensive the parts. Forged pistons will typically cost two to three times as much as cast. A set of eight forged pistons will cost between $200 and $500 or even as much as $1,000.

Design

Type of skirt

There are two basic skirt designs used in the engines of the last few decades. The full skirt or trunk piston was very popular when we all liked Ike. In those days, the crankshafts did not have counterweights. Trunk pistons resemble an inverted can, a cylinder with one end sealed. Many applications in those days put the two compression rings at the top of the piston, where they are found today, and the oil control ring near the bottom of the piston.

The slipper-skirt piston has large notches cut in the bottom. This will allow clearance for the counterweights as they swing past. Counterweighted crankshafts have been used in most engines of the last three decades. The large counterweights serve to smooth the vibration of the engine at all rpm.

Piston crown type

Flat head

The flat head piston is common in stock engines and should not be confused with a flathead engine. When this piston is at top dead center, the top surface of the piston will be level, smooth, and flush with the deck of the block. Most gasoline engines leave the factories with flat head pistons.

Cup head

Cup head pistons are often found in the low-compression, emission-control engines of the 1970s and early '80s. While the perimeter of the piston head is flush with the deck of the block, the center of the piston is recessed. Since part of the combustion chamber is the recess of the piston, the volume of the combustion chamber is large even when the piston is at top dead center, thereby lowering the compression ratio. The advantage to the manufacturer is that lower compression means lower oxides of nitrogen emissions (NOx), and lower NOx make the engine and the vehicle the engine is mounted in more palatable to the Environmental Protection Agency (EPA).

Dome head

When I was in high school I thought that all would be right with the world if only I had a hemi. Now keep in mind, I really did not know what a hemi was; all I knew was that I needed one. The term hemi, of course, refers to the design of the combustion chamber. A hemispherical combustion chamber is shaped like the inside of a domed building. The spark plug is located in the center of the dome and creates a flame front that will theoretically propagate equally in all directions. The result is that the power created by the burning air/fuel mixture is directed toward the top of the piston. The shape of the hemispherical combustion chamber would inherently cause a low-compression ratio if the pistons used in a hemi engine were flat head. A typical engine with hemispherical combustion chambers will use domed pistons to ensure the compression ratio remains high enough to take advantage of the power from the engine.

Hump head

The hump head piston features a raised area designed to partially fill the head's combustion chamber when the piston is at top dead center. Using this type of piston is one of the easiest ways to increase the compression ratio of an engine.

Contour head

Contour head pistons are similar to the hump head pistons. The goal is to fill the combustion chamber, reducing its volume to the smallest amount practical. The smaller the combustion chamber volume when the piston is at top dead center, the higher the compression ratio. The hump head piston has only a rise into the combustion

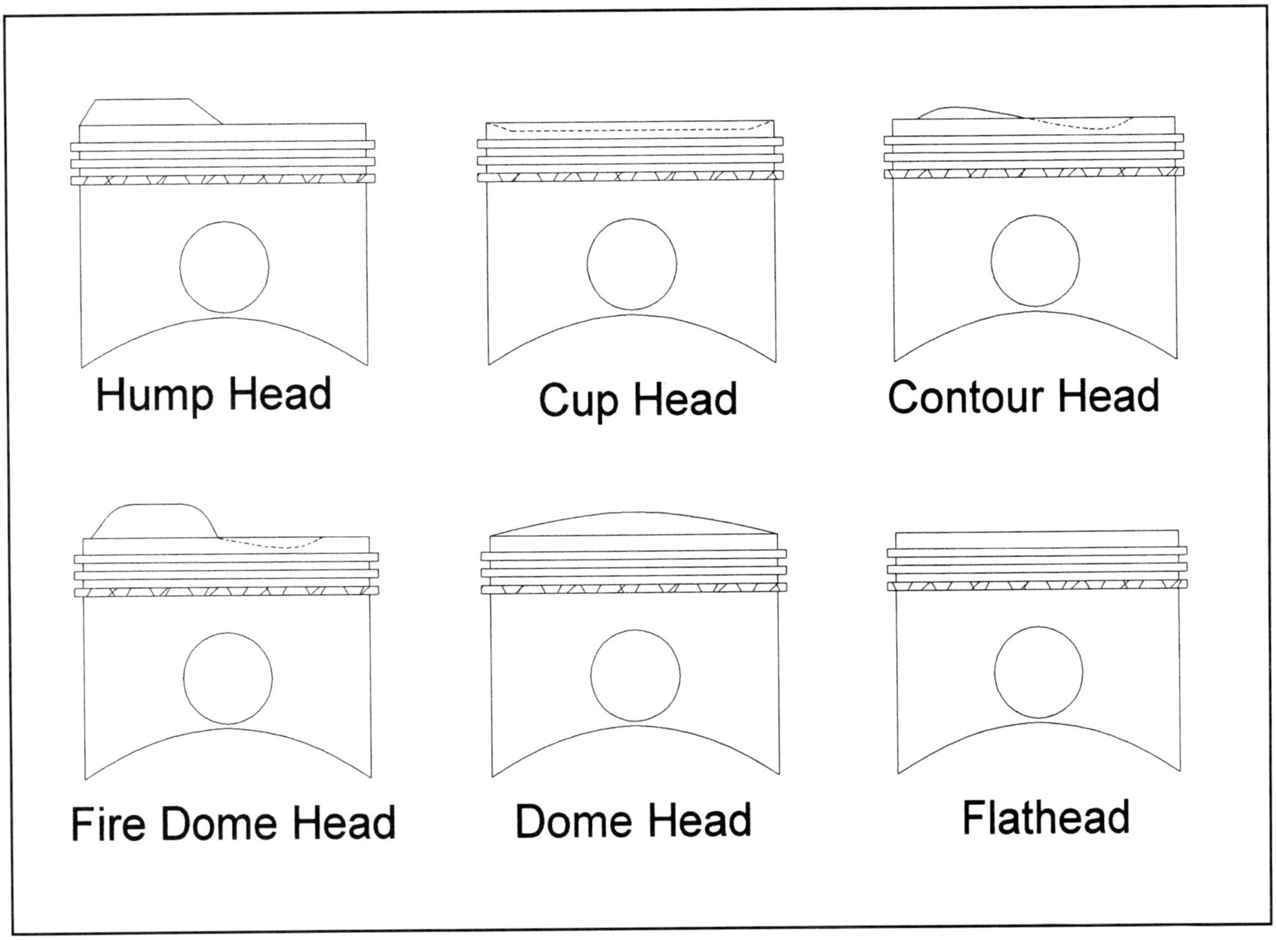

The heads of pistons have a variety of shapes. Each of the shapes is used to accomplish a specific goal. The hump head, fire dome, and contour head pistons are designed to fill the combustion chamber of the cylinder head. The dome head piston is designed to be used with hemispherical combustion chambers. The flat head and cup head pistons are what are typically found in stock engines. The cup head piston is often employed by a manufacturer to lower compression ratios for emissions purposes.

chamber of the cylinder head, while the contour head piston tries to fill the combustion chamber. These pistons are designed specifically to maximize compression ratio. In addition to a rise, the contour piston also has a depression. This depression helps to direct the expanding gases of combustion, keeping them concentrated on the crown of the piston.

Fire dome head

The fire dome piston is a further refinement of the contour head piston. Again, the goal of the shape of the crown of the piston is to direct the expanding gases of combustion, keeping them concentrated on the crown of the piston.

Piston pin offset and thrust surfaces

Most piston wrist pins are offset from the centerline of the piston by several thousandths of an inch. Since the piston is always slightly smaller than the cylinder into which it is installed, the piston has room to move in the cylinder. Since the connecting rod pushes the piston up at an angle during the compression or exhaust stroke, the piston is forced against the opposite wall of the cylinder. At this point, the piston is parallel with the wall on the opposite side of the cylinder from the rising swing of the crankshaft and connecting rod. The piston is having its minor thrust surface pressed against the wall of the cylinder. When the piston reaches top dead center, the connecting rod that had been pressing in one direction will now begin to pull from the opposite direction. Of course, if the cylinder is on the power stroke the expanding gases pushing down on the piston forces it against the opposite cylinder wall. At this point, the major thrust surface is being pressed against the opposite cylinder wall. At the point where the piston changes from being pressed against the minor thrust surface to being pressed against the major thrust surface, the piston will tend to rock. If the rocking action is severe enough, it will cause a sound audible to the technician or even the driver of the car. This sound is called piston slap. The offset wrist pin design reduces the rocking action and therefore the piston slap.

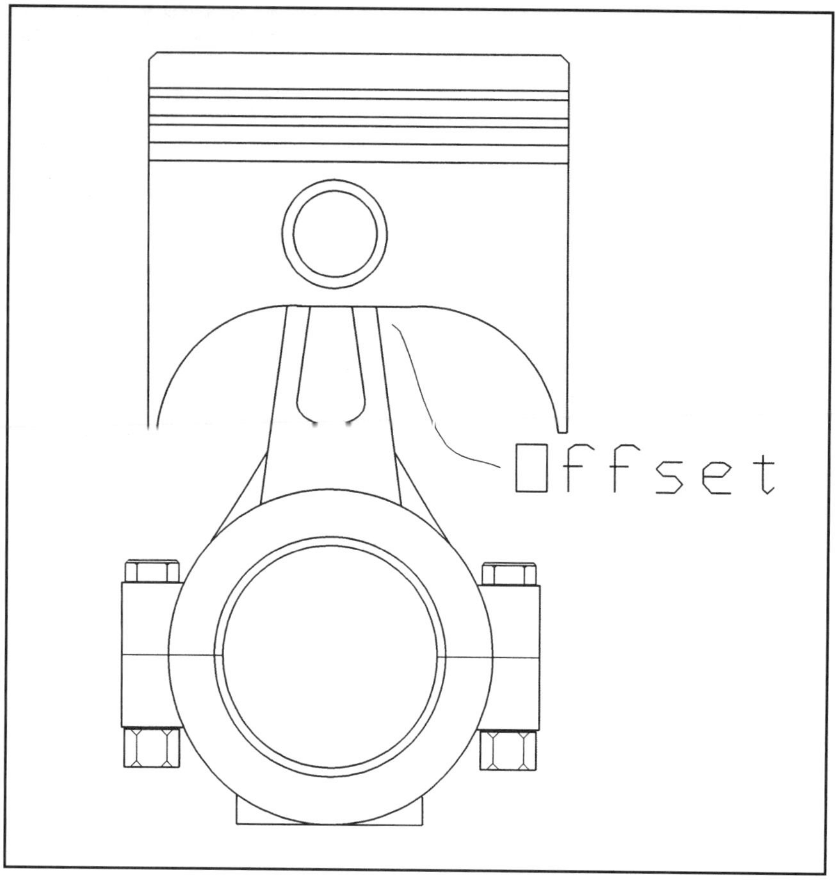

Because the rotation of the crankshaft tends to place more pressure or thrust on one side of the piston than on the other, there tends to be a rocking or tilting of the piston. To redirect or compensate for this tilting, the piston pin is usually offset. In this exaggerated drawing, the offset shortside is the direction of the thrust.

Piston weight

The weight of the piston will have an effect on a number of issues related to the power, maximum speed, and smoothness. In a stock engine the pistons should weigh within a few grams of one another. Now I have always had a hard time figuring out what a gram is. One ounce is about 28.35 grams. Picture this: At the time of this writing the price of a first-class stamp suitable for mailing 1oz was 32 cents. If the price of mailing a letter was 32 cents a gram, it would cost over $9 to mail a 1oz letter. In a performance engine, the entire reciprocating mass—piston, wrist pin, and connecting rod assembly—should be within 0.5 grams of one another.

A piston that is heavier than the rest will create vibration and add stress to the crankshaft. It takes energy to cause a vibration, energy that would otherwise be used to push the vehicle down the road. Vibration will reduce the performance of the engine. One gram of weight difference traveling at 45mph (about 8000rpm), like the typical high-performance piston, has a force of 2.44psi when it changes direction at top dead center and bottom dead center and accelerates toward the midpoint of the stroke. Being off by a whopping

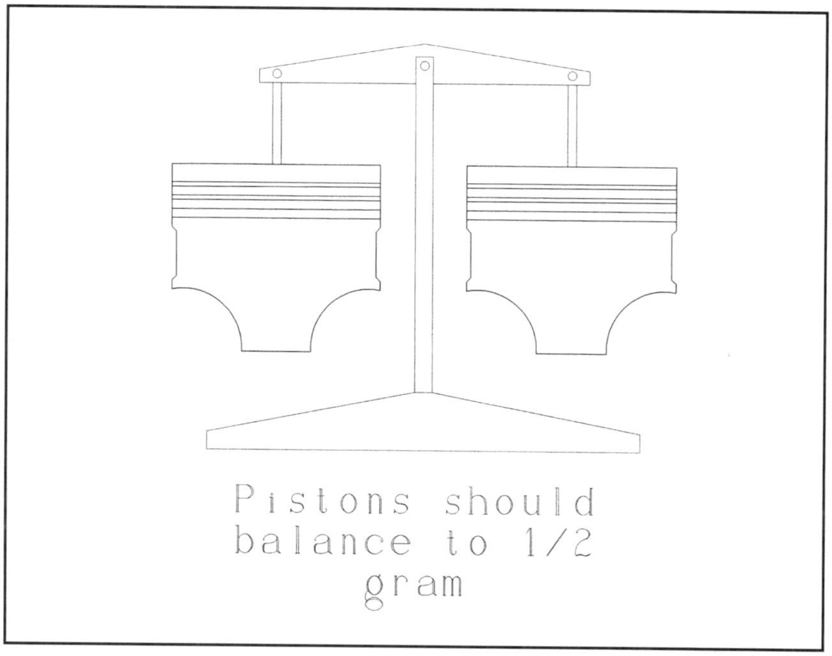

Left
Simply put, lack of balance in the reciprocating components of the engine translates directly to vibration. The weight of a balanced set of pistons should not vary by more than 1/2 gram.

great factor like 1oz would cause the overweight piston to create nearly 76psi of force.

Although great care is generally taken by piston manufacturers to match sets of pistons by weight, they do not often succeed. Frankly, the preceding statement suffers a mild irony. For this reason anyone considering an engine rebuild and considering the replacement of only one piston should reconsider or be prepared to pay a machinist to match the weight of the new piston to the weight of the other pistons. A little later we will discuss how to match the weights of the pistons.

The ideal piston would be made of as light a substance as possible. When accelerating eight pistons with 76psi of force, each consumes a large amount of power. The heavier the pistons, the greater the power consumed by their acceleration and deceleration on every crankshaft rotation.

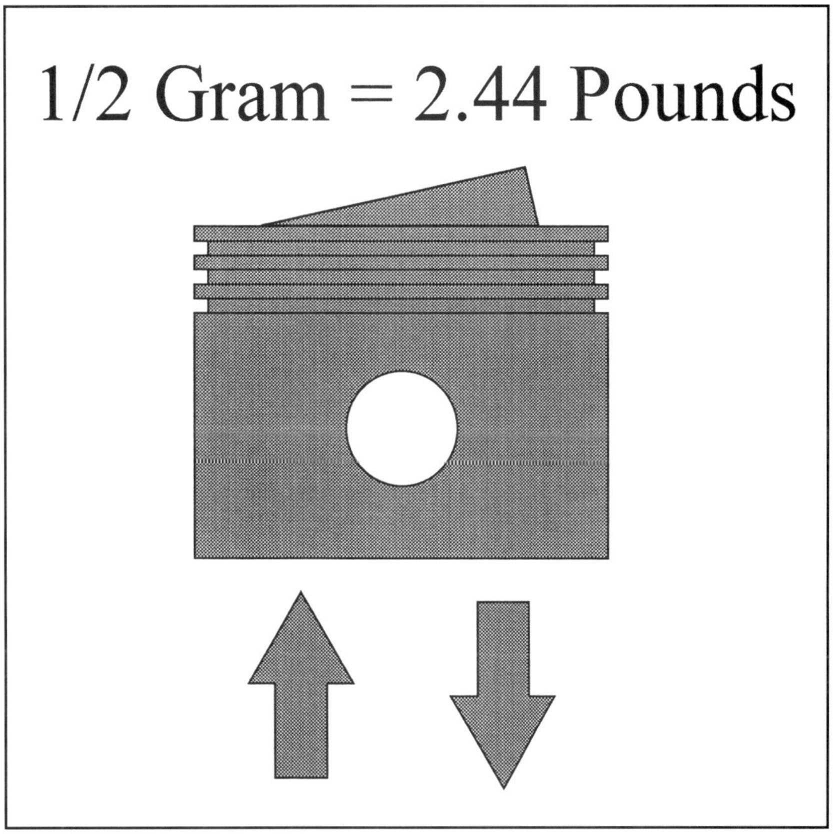

At the point where the piston changes direction, both at top dead center and at bottom dead center, 1/2 gram translates to a 2.44psi difference in force. When each piston has a different influence of force on the crankshaft at both bottom dead center and top dead center, vibration results.

Piston wear— Measuring the pistons

If there is any serious damage to the cylinders, do not bother to measure the pistons because you are going to replace them. If you have determined that the cylinders will need to be rebored, do not bother to measure the pistons because you are going to replace them. If the cylinders are in good condition, the pistons are probably in good condition.

Piston measurements include size, clearance, and taper. Since the important part of the inspection of pistons is visual, actual measurement of the pistons is to confirm their apparent good condition.

Measure the size of the piston 3/4in below the centerline of the piston perpendicular to the wrist pin. Compare this reading to the factory specifications for the engine you are rebuilding.

To measure piston clearance, subtract the maximum piston

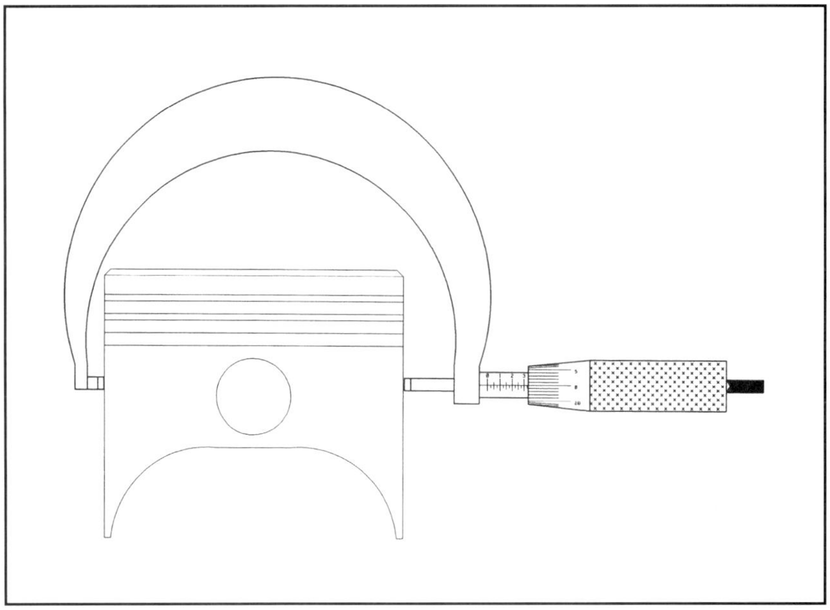

After the pistons have slid up and down the cylinder a few million times, there is a tendency for them to wear. Measuring the pistons should be a routine part of an overhaul. Use some common sense. If the block is going to be bored, the pistons will need to be replaced anyway.

diameter from the diameter of the cylinder. The difference between these measurements is the piston clearance. The clearance for pistons to be reused should always be less than 0.0035in. If the clearance is excessive, consult with your machinist about the advisability of having them knurled. Knurling raises the metal on the skirt of the piston to reduce the piston to cylinder wall clearance.

To find the piston taper, measure the diameter of the piston just below the oil ring and the bottom of the skirt. The difference between these measurements is the taper. Taper should not exceed 0.010in.

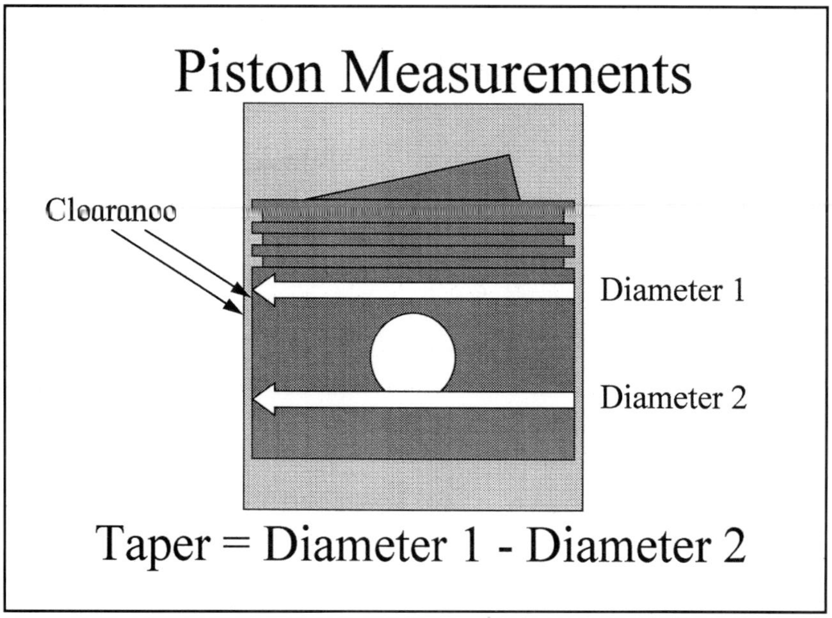

There are three basic measurements of pistons: diameter (size), taper, and clearance. Taper is the difference in the diameter of the piston measured at the top and bottom of the skirt. Clearance is the difference in the diameter of the piston and the diameter of the cylinder.

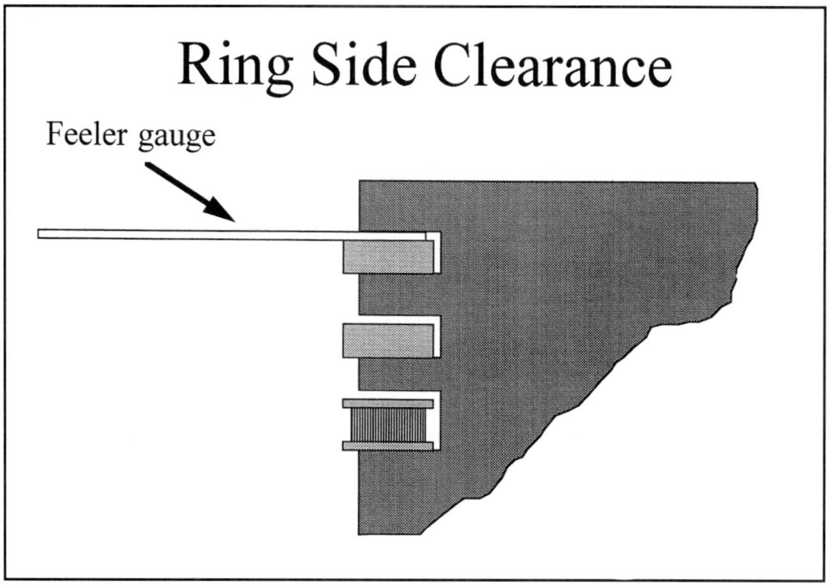

Ring side clearance is a critical measurement that is often ignored by even the most experienced rebuilders. Side clearance is a measurement of how much wider the ring groove is than the ring. Too much side clearance and the ring can twist and break. Too little side clearance and the ring can bind, become stuck, and cause a loss of compression.

Ring groove wear and side clearance

Ring side clearance should also be checked. Place a feeler gauge in each of the ring grooves with the ring destined for that groove installed. Progressively increase the thickness of the feeler gauge until a snug fit is obtained. Ring side clearance should be 0.002 to 0.004in.

Typical overhaul operations for pistons

Cleaning the pistons

When you decide to build up your engine for performance you should make friends with someone who owns a machine shop. Maybe marry his daughter, maybe promise not to marry his daughter. Having a friend with solvent tanks, caustic hot tanks, bead blasters, and personnel who you can con into doing the work for you can be very beneficial to the progress and cost of your project.

The first step to cleaning the pistons is to clean the top of the pistons with a wire wheel. As you are doing this, imagine the top of the piston to be the hood of your new red Dodge Viper. Imagine the residue on the top of the piston to be a light layer of dust. Imagine the wire wheel to be a buffer. If you were to use it carefully and gently it would remove the dust with a minimum of damage to the finish of the Viper. On the other hand, harsh use of the buffer would convert the smooth finish into a red-on-red paisley. A wire wheel turns at an extremely high speed; if you are not careful, severe damage or wear can be caused to the head of the piston. If you do not have access to a wire wheel, a scraper, gently applied, may be used to clean the residue off the top of the piston.

Once the head of the piston has been buffed clean, it is time to find the use of a bead blaster. Protect the wrist pin bores. Fine grain-sized corks will do a good job. Gaffer's tape around the inside of the bore will also do well. Gaffer's tape can be purchased from your local professional film production rental house. It is tough like duct tape but does not leave behind a sticky, gooey residue. Masking tape will probably be destroyed during the bead blasting and therefore fail

Ring groove side clearance should only be measured after the ring grooves have been cleaned and preferably with the rings you plan to use at final assembly.

There are varied opinions about cleaning the pistons with a wire wheel. Remember that the head of the piston is delicate and can be damaged by the wire wheel. Use the wire wheel gently. The lands and grooves of the piston should never be cleaned with a wire wheel.

to offer sufficient protection. Bead blasting is intended to clean the ring grooves and skirts of the piston; it is not intended to reduce the mass of the piston. Excessive blasting can damage the ring grooves. Damaged ring grooves can result in compression leaks. In a precision performance engine, the ring grooves should be protected like the wrist pin bores. Actually, bead blasting a performance engine is to admit that you are willing to sacrifice performance for a few hundred dollars in cash. If the performance engine pistons are in bad enough condition to warrant bead blasting, they are in bad enough condition to warrant replacing.

Note: Do not use a wire wheel on the skirt or ring grooves of the piston.

Cleaning the ring grooves

Be sure that the ring grooves are clean and clear of debris and carbon. Use a special tool called a ring groove cleaner. For those

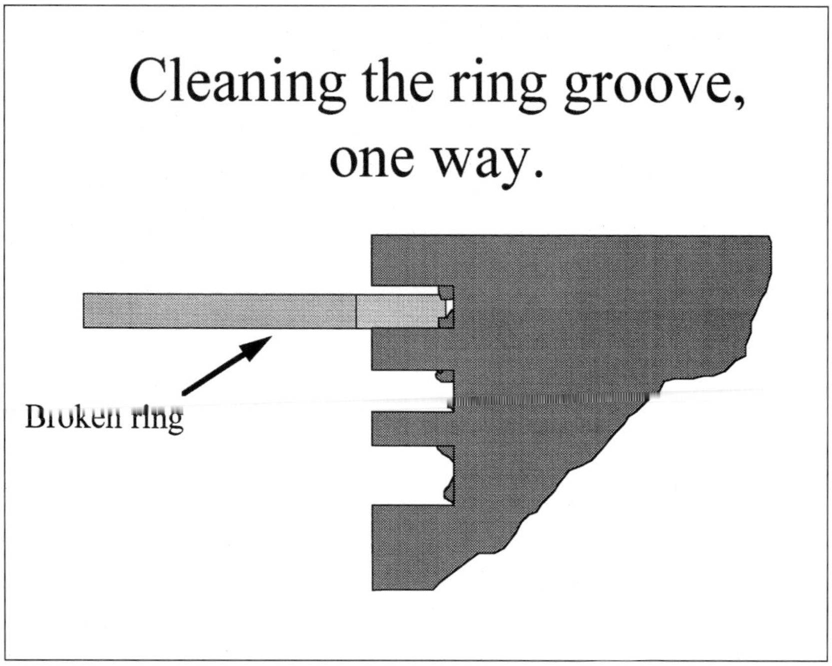

There are special tools sold to clean the ring grooves. But like many special tools I own, it stays in the toolbox while I revert back to a more familiar technique. A pair of thick gloves and a broken compression ring can clean the groove very nicely. Gloves are essential because the broken ring is sharp and can cause nasty lacerations. Remember that when cleaning the ring grooves the idea is to clean the carbon out of the grooves, not to machine the grooves deeper. If you start seeing shiny metal in the material you are removing, then reduce the pressure you are using.

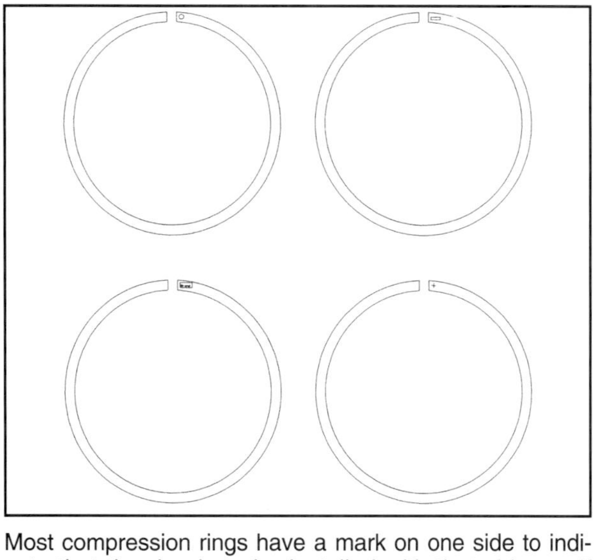

Most compression rings have a mark on one side to indicate that the ring is to be installed with that side up. If there is no mark on the ring, then refer to the installation instructions provided in the package.

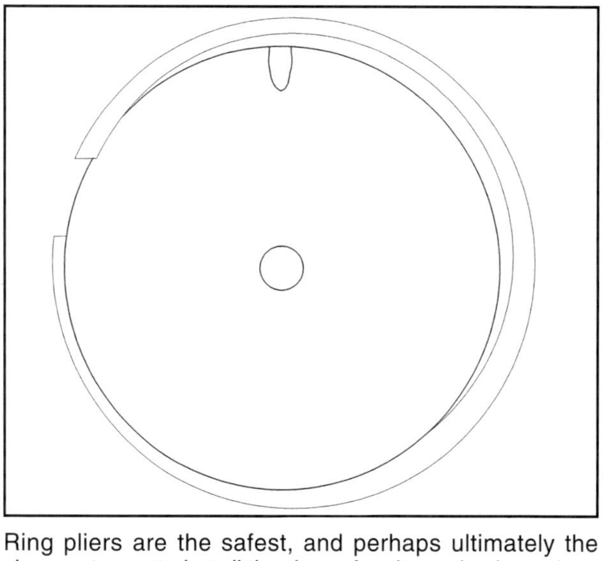

Ring pliers are the safest, and perhaps ultimately the cheapest, way to install the rings. An alternative is to place one end of the ring in the ring groove of the piston and gently rotate the ring around the piston until it falls into the groove. I've used this technique for many years and still occasionally break a ring. Replacing that ring, especially if they only come in full sets for your engine, can be very expensive. Use piston ring pliers.

who have good health insurance and a first aid kit, a broken ring can be used to scrape the carbon from the grooves. Extreme caution should be used when using a ring to clean the grooves. The broken ring is sharp and can leave nasty cuts on the hands and other anatomical parts. Wear heavy gloves.

With the ring grooves clean, install the new rings on the pistons. Stagger the ring gaps before installing the pistons into the cylinder. Staggering will provide for a better seal during the early stages of running the engine. Keep in mind that on most pistons the rings will rotate as the engine runs. Do not be shocked if the rings are all lined up when you disassemble the engine again in a few days, weeks, or years. Also do not be surprised if they are not lined up. The rings are usually identified top, middle, and oil control by the papers covering them in the box. There will also be a mark on the two compression rings indicating which way is up. This mark may be a dot, a bar, or even a corporate logo. In a few instances, the rings may be unmarked, there will be instructions in the box.

Two techniques are available for the actual installation of the rings. Start one end of the ring in the correct groove. Rotate the ring around the center of the piston and it will practically fall into place. Plan on breaking an occasional ring using this technique. Piston ring installation pliers can be purchased for only a few dollars at most auto parts stores. Using these pliers virtually eliminates the possibility of breaking rings.

Knurling the pistons

Knurling the pistons is one of those "cheap fixes" for a problem that should demand replacement of the pistons. Knurling the pistons can only be used to compensate for worn piston skirts. There may be times when appropriate, but generally, if in doubt, replace the pistons in question. You have better things to do than rebuild this engine again next year.

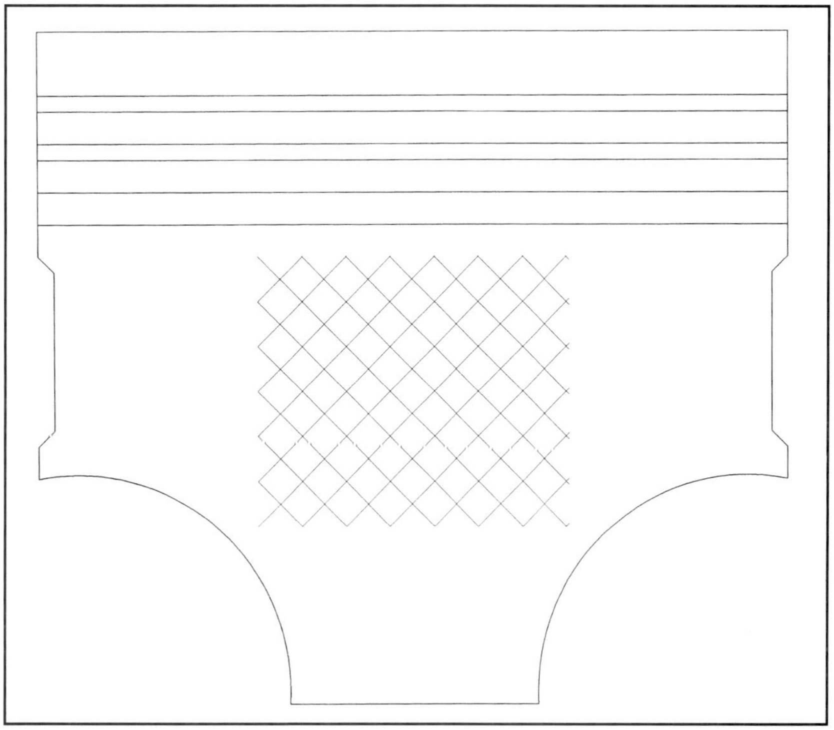

Knurling the pistons is one of those "cheap fixes" for a problem that should demand replacement of the pistons. Knurling the pistons can only be used to compensate for worn piston skirts. There may be times when it is appropriate, but generally if in doubt, replace the pistons in question. You have better things to do than rebuild this engine again next year.

If you choose to have the pistons knurled, scrape the knurling with an old wrist pin before you install them. Slide the wrist pin vertically from the bottom ring groove to the end of the piston skirt. This ensures the rough edges of the knurling have been removed.

Okay, now that you know about the process of knurling pistons, let us get a little more realistic about it. You probably bought this book with the idea that you would learn about performance engine building. Putting knurled pistons into a performance engine is like serving a McDonald's hamburger on Wedgwood china.

Blueprinting and the pistons (not so typical overhaul operations for pistons)

Piston-to-wall clearance

In the sauna of the summer of 1971, I worked pit security at the Dallas Springnationals. Now, at my tender age, that was a real experience. I actually got to stand within 100 yards of Linda Vaughn, a woman who had dated Jim Clark! Anyway, this was a meet featuring "The Snake," "The Mongoose," Don Garlits, and anyone who was anyone in the business or pastime of drag racing. In spite of this, the car I remember the most was a gasser rail that belched smoke until he revved the engine up. When I inquired about this, I was told that the engine had been "set up loose" to reduce the friction of the pistons in the cylinders. Even through the fog of Watergate, disco, the

John Lennon assassination, Reaganomics, Desert Storm, and failed health care reform, I am still trying to figure out if the guy was kidding me.

Excessive piston-to-cylinder wall clearance will allow the piston to rock excessively as previously described. If the clearance is insufficient, the piston can literally seize in the bore. Keep in mind that the piston is cam ground, or oval shaped. As most substances will, the piston will expand when heated. Since the structure of the piston is not concentric, the amount of expansion will be different on the wrist pin axis and 90 degrees to the wrist pin axis. The oval shape when cold will allow the piston to be round when the engine is warmed up. The difference, when cold, between these two axes is only about 0.020 to 0.040in.

Most pistons are also tapered 0.005 to 0.010in. The bottom of the piston skirt has less metal to expand than does the area closest to the head. For that reason, the top of most piston skirts is smaller than the bottom of the piston skirt. There are also pistons that have a barrel grind. These piston skirts are narrow on the top and bottom and wide in the middle.

Another factor is the material the piston is made of. If you have enough money to make Ross Perot feel like a pauper, you might invest in ceramic pistons. These may not expand at all; the far more common aluminum pistons will have expansion rates that will vary depending on the alloy's blend of silicon, magnesium, manganese, nickel, copper, and other elements.

The last three paragraphs lead to this simple fact. To ensure proper piston-to-cylinder wall clearance, the diameter of the piston must be measured at the point recommended by the manufacturer of the piston. Further, the clearance should be that specified by the piston manufacturer not by the factory service manual.

Begin the process of checking piston-to-cylinder wall clearance by reviewing the measurement recommendations supplied by the piston manufacturer. This recommended measurement will often be 90 degrees to the wrist pin at the top of the skirt (don't count on it). Measure this diameter carefully with a micrometer. Now with an inside micrometer, mic the diameter of the cylinder at several places along the length of the cylinder. The places checked should be on the same axis as the piston manufacturer recommended for checking piston diameter. Further, the places checked should be in the area of travel of the mic point of the piston.

A simpler, though often just as effective, method to measure piston-to-cylinder wall clearance is with a feeler gauge. The best way to use the feeler gauge for this method is the go-no-go method. Read the piston manufacturer's recommendations for clearance. Select a feeler gauge that is equal to the minimum recommendation and a feeler gauge just larger than the maximum recommendation. The thinner feeler gauge should slide easily between the cylinder wall and the piston, while the thicker gauge should not fit at all.

Whatever the method you use, be sure to make these measurements exact. Remember that failure to follow the piston manufacturer's recommendations exactly can cause piston slap or seizing of the piston.

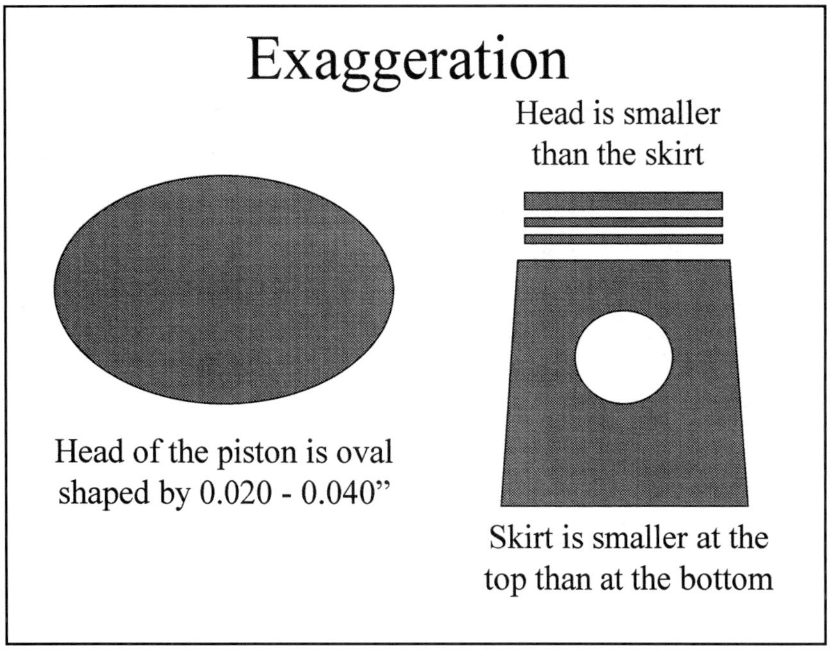

The piston will tend to expand as it heats up. The thickness of the piston varies from place to place. Therefore, pistons will not be round. Each axis will expand at a different rate that is dependent on the amount of metal on that axis.

Track performance pistons and the piston-to-cylinder wall clearance

An engine that is destined for the world of mufflerless open headers on the drag strip has different requirements from one that is designed to haul ma and the kids to the local K-Mart. On these performance engines, the sound of the exhaust will muffle any piston slap that might occur. For these engines it is recommended that the piston-to-cylinder wall clearance be exactly at the maximum limit. The larger clearance will prevent piston scuffing due to high heat and pressures when the tachometer begins to make its second lap during competition.

Honing for exact cylinder diameter

Proper honing of the cylin-

Below
Honing machines like this can be used to precisely establish the right cylinder bore diameter. Even a freshly bored engine may have some anomalies that can be corrected through honing. Do not confuse this process with the hand honing that is done to establish the cylinder wall crosshatch, however. Honing the cylinders to alter bore anomalies can only be done by enthusiasts who have no other life. It is a time-consuming process to perform by hand.

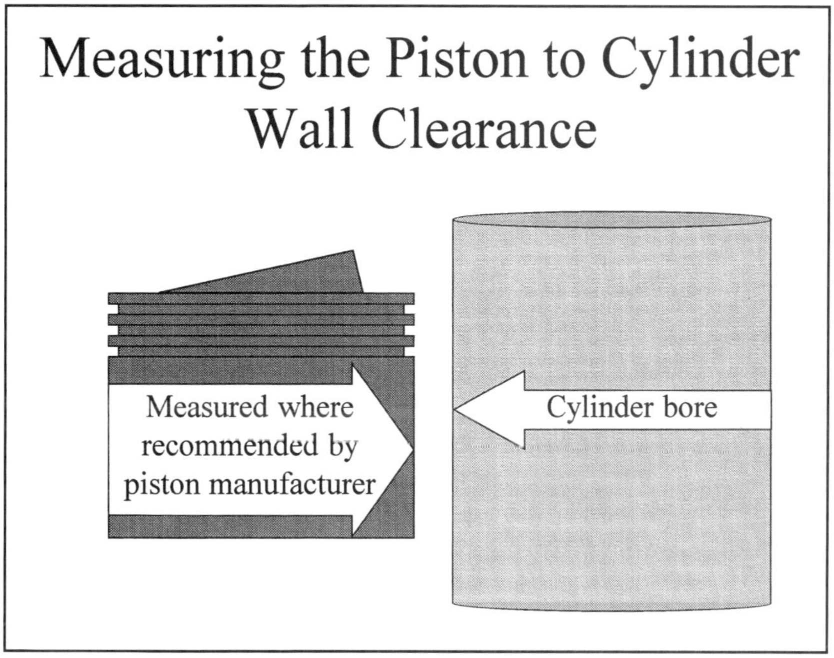

Piston-to-cylinder wall clearance is a measurement of the diameter of the cylinder less the piston diameter. Remember that only the piston manufacturer knows the expansion characteristics of its pistons. When measuring the piston diameter, be sure to follow the manufacturer's recommendations.

ders can ensure that the cylinder bore is exactly the correct diameter for each piston. Correct honing technique will produce a cross-hatch pattern on the cylinder walls. Only a flat stone hone can be used to true the cylinder walls. Although a flat stone hone mounted in a drill motor will do a reasonably good job for a stock engine, maximum precision for performance can only be achieved with your machinist's honing machine.

There is a "home" version of a precision honing machine that does an excellent job. If you have a lot of time and a lot of money, or if you are going to be building a lot of engines, then investing in a "home" honing machine might be worthwhile. Proper performance honing requires that the block be submerged or constantly bathed in oil during the process. Along with the honing machine, buy an old bathtub and several gallons of honing oil. Each cylinder should have a determined ideal bore. The cylinder should be honed for a minute or so, then measured. This may need to be repeated dozens of times. Also, since there is likely to be 0.001 to 0.006in of metal to remove from each cylinder, be sure to verify that there is nothing you want to see on television for the next week.

If you are having the block bored, it is standard procedure for the machinist to then hone the cylinders. The boring machine will bore the cylinders 0.004 to 0.006in undersize. Honing is then used to remove the remaining metal for proper bore size. It would not hurt to confirm that honing will be part of the procedure when you drop the block off with the machinist.

Deburring

Anyone who has built a model car or model airplane is familiar with the flashing that results from seepage of the plastic into the joints between the sections of the mold. When building a plastic model, this flashing must be removed to make the final product look neat.

In a piston, there is flashing that marks the junction between the pieces of the mold and at the edges of the piston. Removal of this flashing is for more than purely cosmetic purposes. The ridges formed by the flashing provides a starting point for cracks and stress fractures. Most high-quality performance pistons should require very little deburring. Unlike deburring an engine block, all deburring of pistons should be done by hand using a moderate grit emery paper or a flat file.

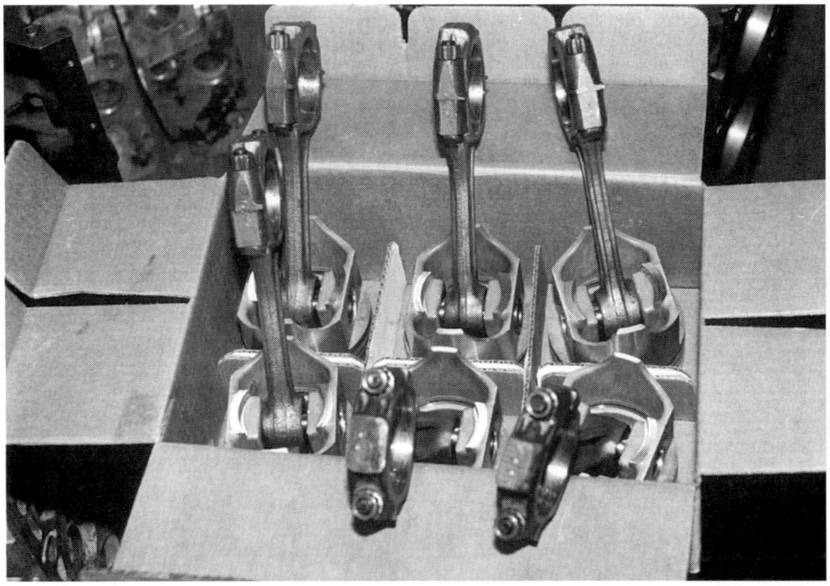

This is a fresh set of pistons newly installed on "experienced" connecting rods. All ready to be installed, right? Closer inspection would reveal sharp ridges left behind from their casting.

These sharp casting ridges provide a place where high stresses can be concentrated over a small area. Bottom line: These are places where cracks and breaks can begin.

Ring grooves

Of all the parts of a piston, probably the parts that require the most accuracy and precision are the ring grooves. The metal area between the grooves is known as the lands. The head of the piston and the lands are several thousandths of an inch smaller in diameter than the rest of the piston. The bottom and top surfaces of the lands must be as smooth and parallel as possible. Any roughness or uneven surface on the lands can cause compression leaks and decrease power. The piston ring lands should be protected like negotiable stock certificates. A small ding or warpage of a land on one can literally render that piston, and therefore possibly a set of pistons, useless. If the ring grooves are carboned up on a street engine, you scrape the carbon using a special tool or with a broken ring. Performance pistons with carboned grooves should just be replaced. Of course, if the engine is being used on the track, it probably will not get enough mileage put on it to get carboned. Until the pistons are ready to be installed in the engine, keep the lands and grooves protected with gaffer's tape or something similar. This is crucial on the performance engine piston if the engine is destined for competition use. Building a performance engine is definitely a job for the detail-oriented individual.

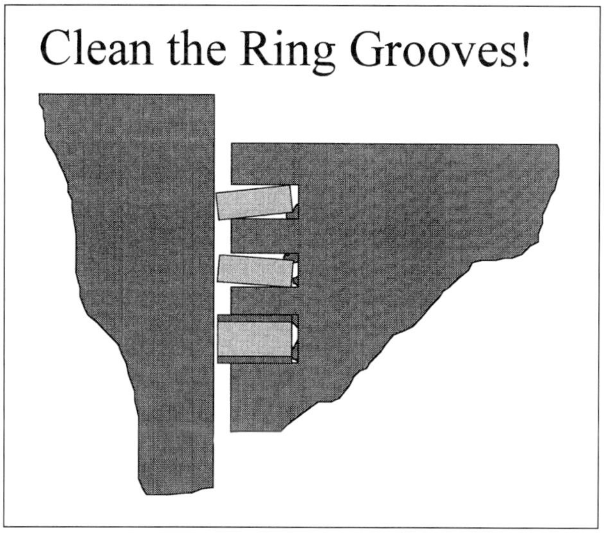

If the ring grooves are not properly cleaned, the dirt and particles in the grooves can put extreme torsional forces on the rings during combustion. This is not only important on performance engines, but it is also important on engines that are going to be expected to last for years and thousands of miles.

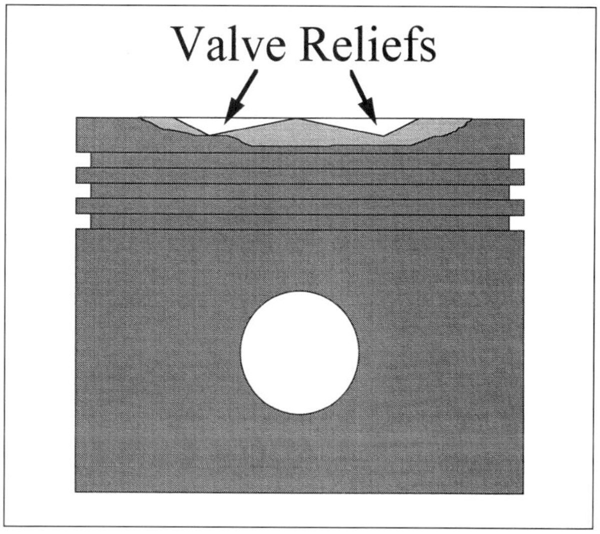

Many pistons have notches in the head. These notches provide space in the tops of the pistons for engines that have a lot of valve overlap. In the event that the piston is still near top dead center as the intake valve opens, clearance is provided.

Valve reliefs

Now, if you were going to design a spark-ignition, internal-combustion engine you might be inclined to take the simplest approach possible. The piston needs to be at the top of its stroke when the intake valve opens. This way when the piston begins its downward journey, the intake valve will be open and an air/fuel charge can fill the cylinder. The problem is that the intake valve must be at least partially open when the piston arrives at top dead center to ensure that the cylinder can fill properly with air. The time that is most critical, in fact, is when the piston is accelerating down the first half of the stroke. The valve must be fully open before the piston reaches its point of maximum velocity at the halfway point of the stroke. Often, and especially at high speed, the intake valve will need to be partially open when the piston arrives at top dead center. In a low-compression engine there is usually plenty of clearance between the factory valve reliefs in the top of the piston and the bottom of the intake valve. In a high-compression engine, it may be necessary to machine or modify the factory valve reliefs.

Piston-to-head clearance

The Mustang engine described earlier had been assembled with a negative piston-to-head clearance of nearly 0.050in. This reduced the power of the engine, to say the least.

Measuring piston-to-head clearance is another of those modeling clay tricks. Lightly coat the surface of the piston and combustion chamber with a thin oil. Place a small ball of modeling clay in the center of the piston. Make sure the ball is large enough that it will be compressed between the top of the piston and the bottom of the head. Also make sure that the ball is not so large that the piston cannot go over top dead center. Install the cylinder head with the type of gasket that you will ultimately use during final assembly. Torque the cylinder head. Rotate the crankshaft through 360 degrees of rotation. The minimum thickness of the modeling clay will be your minimum piston-to-head clearance.

Piston-to-valve clearance

If, as previously mentioned, you have a bank account that would make Old Man Rockefeller faint, simply install the high-performance pistons, start the engine, warm it up, and rev it to 7000 to 9000rpm. If the engine holds together, then there is sufficient piston-to-valve clearance. For most of us, this may not be practical.

Oddly enough, piston-to-valve clearance is less critical on a high-performance track engine than it is on a street performance engine. These true racing engines feature valvetrains that are very positive in action. The valve springs are extremely stiff to prevent valve float at high engine speeds. These valve springs may be so stiff that only a few hours of operation could do severe damage to the camshaft. Many feature roller lifters to reduce wear. Street performance engines must be built with the idea in mind that the engine will

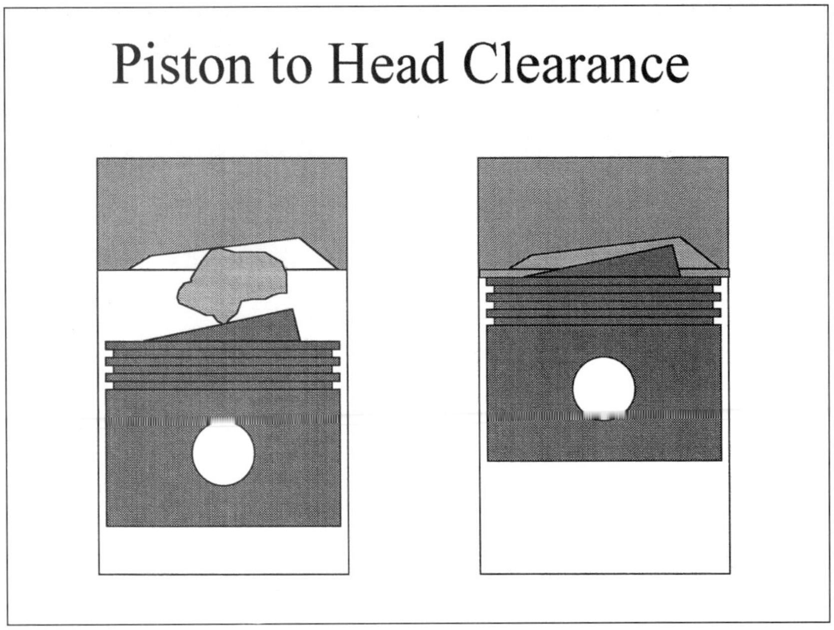

Place a small ball of modeling clay in the center of the piston. Make sure the ball is large enough that it will be compressed between the top of the piston and the bottom of the head. Also make sure that the ball is not so large that the piston cannot go over top dead center. Install the cylinder head with the type of gasket that you will ultimately use during final assembly. Torque the cylinder head. Rotate the crankshaft through 360 degrees of rotation. The minimum thickness of the modeling clay will be your minimum piston-to-head clearance.

To get maximum performance from an engine, it is necessary to precisely synchronize the camshaft and the crankshaft. This process is called phasing. It is necessary to phase the camshaft when the engine is finally built; therefore, it is necessary to phase the camshaft when checking the piston-to-valve clearance. Shortcutting this process can yield perfect results when just testing the clearance and a disastrous lack of clearance when the engine is finally assembled. Attach a degree wheel to the front of the crankshaft.

last for more than one trip to Sears. The valvetrains are not as rigid; therefore, at high engine speeds, the valves may begin to float, or remain open even though the lifter is well on its way toward the heel of the cam lobe. This may result in the valve remaining in the same space in the universe that the piston will soon try to occupy. This is a disastrous and deadly event and has often been responsible for scattering pieces of metal all over the inside of the engine and all over the ground.

A not very sexy but practical way to ensure proper clearance is to mold a 1/8in thick layer of modeling clay to the top of the piston. Sprinkle a little talcum powder on the top of the clay to keep it from sticking to the head and valves. This should be done on all pistons to ensure that they will all be attended to for proper clearance. For this method to be accurate, everything must be as it will be when the engine is built up. The pistons must be installed on the connecting rods that will be used, they must be torqued on the crankshaft, and the crankshaft main bearing must be torqued. The type of head gasket that will be used in final assembly must be used.

To get maximum performance from an engine, it is necessary to precisely synchronize the camshaft and the crankshaft. This process is called phasing. It is necessary to phase the camshaft when the engine is finally built; therefore, it is necessary to phase the camshaft when checking the piston-to-valve clearance. Shortcutting this process can yield perfect results when just testing the clearance and a disastrous lack of clearance when the engine is finally assembled. Attach a degree wheel to the front of the crankshaft. Rotate the crankshaft until the camshaft pushes the intake lifter for the number one cylinder to its highest point. The point of maximum lift should not be monitored visually; for accuracy, it should be checked with a dial indicator. Using the recommendations of the camshaft grinder, loosen the bolts in the previously slotted cam sprocket holes and align the cam sprocket so that the intake centerline matches the recommended position on the crankshaft.

With all components installed, phased, and torqued in place, bar the engine over a few times. Remove the cylinder head; the cross-section of the clay is the clearance. Now the problem is judging how much clearance is enough and correcting incorrect clearance.

Sufficient clearance must take into account valve float, connecting rod stretch, and thermal expansion of all components. The general rule is there must be at least 0.080in of clearance

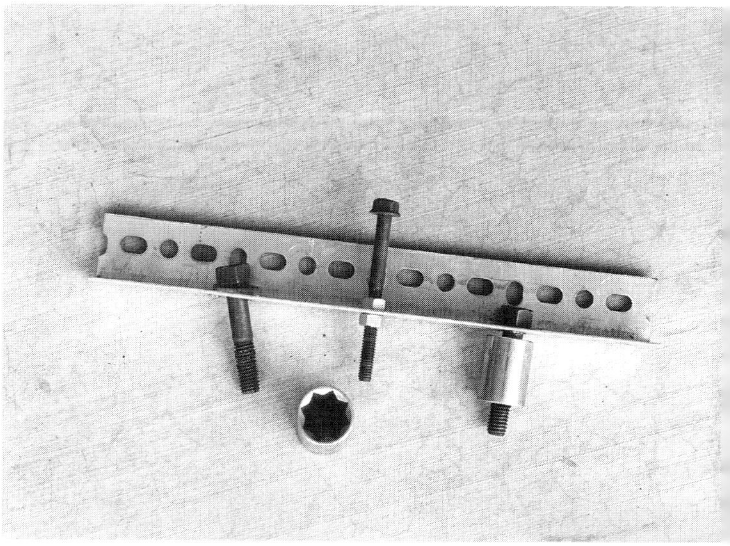

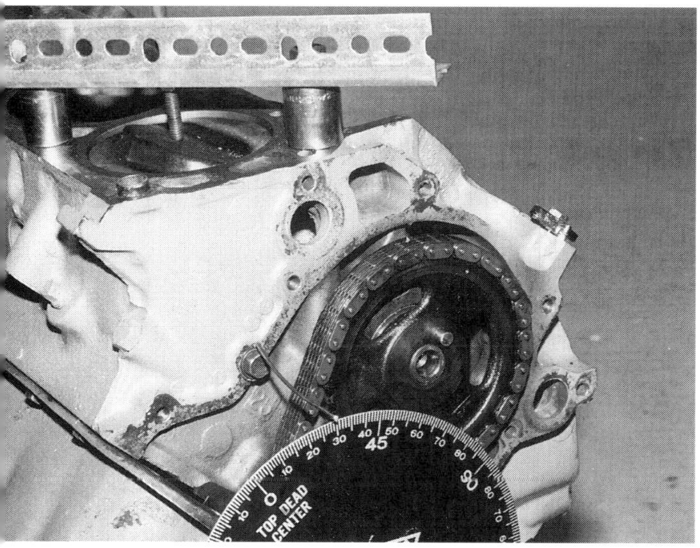

Locate the exact top dead center using a piston stop tool. This one was made from a piece of angle iron, a couple of old head bolts, and a couple of sockets. The dial indicator is only used to ensure accuracy. Rotate the crankshaft in one direction until the number one piston rests firmly against the piston stop. Note the position of the degree wheel. Now rotate the crankshaft in the other direction and note the position on the degree wheel. Top dead center is located exactly halfway between the two noted positions.

Rotate the crankshaft until the camshaft pushes the intake lifter for the number one cylinder to its highest point. The point of maximum lift should not be monitored visually; for accuracy, it should be checked with a dial indicator.

Using the recommendations of the camshaft grinder, loosen the bolts in the previously slotted cam sprocket holes and align the cam sprocket so that the intake centerline matches the recommended position on the crankshaft.

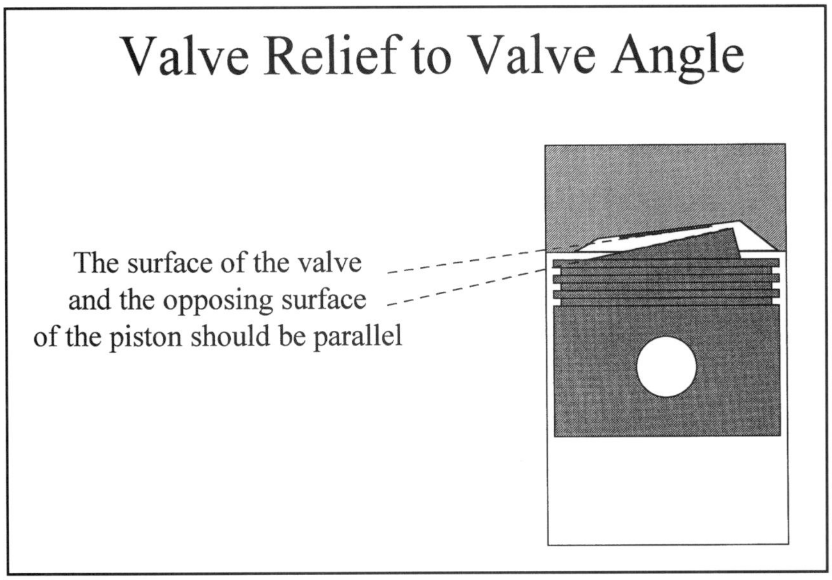

The angle of the valve relief should be the same as the angle of the valve head. This can be measured with a bevel protractor. Measure both the valve relief angle of the piston and the valve head angle. This correction procedure will require the machinist to grind the angle of the piston valve relief area to match the angle of the valves.

for the intake valves and 0.100in for the exhaust valves.

Insufficient clearance

If there is insufficient clearance, the real challenge begins. You can try advancing or retarding the cam phasing. This may, however, be at the sacrifice of performance. The valve seats can be reground; this, however, will affect airflow and therefore power. The best solution is to deepen the valve pockets. This is where you really will wish you had a father-in-law with a machine shop. Deepening the valve pockets is beyond the skills of the typical human, well at least the skills of the vertical mill and flycutter required to do so. The disadvantage (and there is always a disadvantage it seems) of this is that the compression ratio will be lowered a little. A flycutter is a cutting tool with a single bit that rotates around an axis. Racing engines are often plunge cut. Plunge cutting removes only the metal from the valve relief that is absolutely necessary to provide clearance. This will leave a crescent of raised metal. Compression ratio loss is kept to a minimum.

Matching valve relief to valve angle

The angle of the valve relief should be the same as the angle of the valve head. This can be measured with a bevel protractor. Measure both the valve relief angle of the piston and the valve head angle. This correction procedure will require the machinist to grind the angle of the piston valve relief area to match the angle of the valves. This is usually only necessary in engines where the performance is to maximize and where heads have been angle milled. When cylinder heads are milled to an extreme amount for performance, and to maximize compression, the edge of the intake valve seat might be machined away. To prevent this problem, many engine builders may cut mill the head at an angle. This angle is usually no more than 0.125in between the intake and exhaust sides. In truly radical cases, some big block heads are milled at an angle giving up to 0.250in difference between intake and exhaust sides of the head. These radical cases will require the dome of the piston head to be ground at a matching angle.

Dome fitting and radiusing

I recently visited Anderson Air Force Base on the island of Guam. The fellow I visited the base with had been stationed there during the Vietnam War. I commented on a picture of the base that had been taken during the late 1960s. The photograph

If there is insufficient clearance between the top of the piston and the bottom of the cylinder head, then the real craft of performance engine building begins. Flycutters and vertical grinders will need to be used on expensive components to provide the proper clearance.

showed a dozen or more B-52s parked in a straight line. I asked him if the straight line was to make it more convenient for enemy aircraft or saboteurs to destroy a maximum number of aircraft in a single attack. He pointed to dirt mounds that had been built up between the aircraft. He said that the purpose of these mounds was to inhibit fire from spreading from one aircraft to another.

The dome of a piston will act very much like these dirt mounds. The ideal piston would not have a dome. The dome inhibits the progression of the flame front through the combustion chamber. Avoid the use of domed pistons if at all possible. Street engines can maximize power and performance with improved intake and exhaust flow more than they can with increased compression and domed pistons. Some fuels, especially high-octane gasolines (try finding that these days at the corner gas station), methanol, ethanol, and natural gas, give markedly better performance with higher compression ratios. If the engine is destined for any of these uses, you might consider domed pistons. (If you want to know more about alternative fuels, see the final chapter of this book.) However, a street machine running on today's pump gases may detonate so badly that any theoretical improvement will be lost in the reality of the pump gas of the 1990s.

This cylinder head has been previously bolted in place on the engine block. A fine scribe was then run around the circumference of the cylinder on the head. This leaves a circle on the bottom of the head that conforms to where the piston will come closest to the bottom of the head. The piston should fit comfortably and level inside this circle.

This is a set of pistons I installed in an engine over twenty years ago. There was no effort made to ensure they fit properly into the combustion chambers of the cylinder heads. Although there was no contact made with the combustion chamber or valves, the pistons did hammer the spark plugs closed at high engine rpms. At high rpms, the connecting rod would stretch just enough between the intake and exhaust stroke to close the spark plug gap.

Marking the cylinder positions on the head

Begin with a machinist's dye. Color the bottom of the cylinder head and bolt them in place. Mark the circumference of the cylinder on the head with a sharp scribe. Remove the heads and place the piston destined to be filling each combustion chamber in its destined combustion chamber. Take extreme care to ensure that the piston is precisely within the circle made with the scribe. The next step is to make sure that the piston will be positioned with the proper rotation in the combustion chamber. The centerline of the wrist pin hole must have precisely the same alignment when it is positioned in the combustion chamber as it will when the engine is all bolted together. A quick and dirty way to ensure proper rotational alignment of the pistons is to slide a wrist pin in the adjacent pistons. The dome of the piston should comfortably fill the combustion chamber with interference with any part of the head or valves. If the piston domes do fit comfortably, mold a layer of modeling clay over the top of the piston and press them firmly into the combustion chamber, being careful that the piston is precisely positioned within the scribe marks. As discussed before, the clearance at all points in the combustion chamber should be at least 0.080in for the intake valves and 0.100in for the exhaust valves. As hokey as this method may sound, it works quite well on many engines.

For those who were born with a silver slide rule in their mouths, there is a more scientific, though doubtfully more accurate, method of achieving proper rotational alignment of the piston. Set the piston on a machined surface plate. The wrist pin holes of the piston should be held vertically by a vertically mounted pin the same diameter as the wrist pin. Measure the diameter of the piston head. Place a mark at the point on the side of the piston equal to exactly half the diameter of the piston. This will be a point 90 degrees from the baseline established by the wrist pin. Now the head must be marked at a point 90 degrees from the baseline established by the wrist pin. With both the

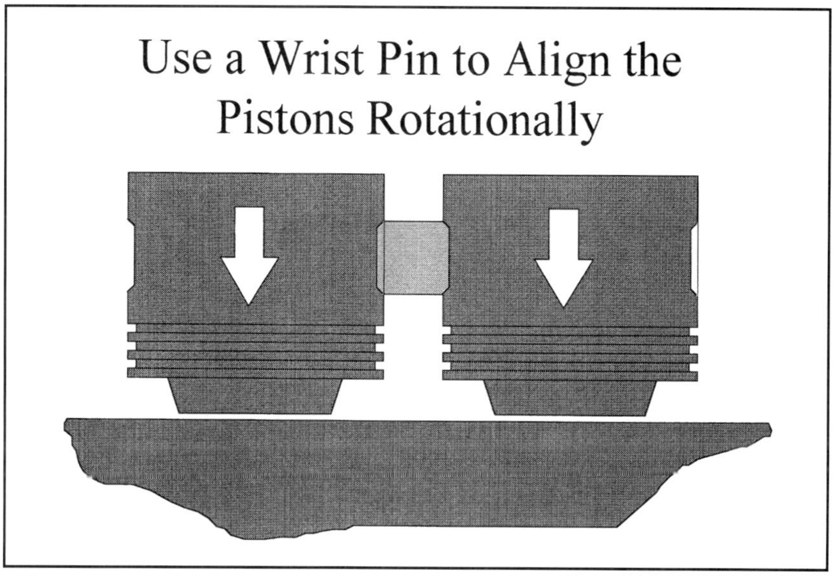

Make sure that the piston will be positioned with the proper rotation in the combustion chamber. The centerline of the wrist pin hole must have precisely the same alignment when it is positioned in the combustion chamber as it will when the engine is all bolted together. A quick and dirty way to ensure proper rotational alignment of the pistons is to slide a wrist pin in the adjacent pistons.

Fitting the piston dome can be an emotion-filled experience. You may have recently laid out as much as $2,000 for pieces of metal that you are about to grind on. Quality tools should be employed. A high-horsepower, high-speed grinder is recommended.

Whether you choose to remove metal from the cylinder head or pistons to achieve proper clearance, remember that every flake of metal removed reduces the compression ratio. This is not usually a bad thing in itself. Most part-time engine builders tend to put too much emphasis on compression ratio anyway. The potential problem is an imbalance in the compression ratios between cylinders. Attempt to remove the same amount of metal from each of the combustion chambers or pistons. After this operation is completed, check the compression ratios of each cylinder using the procedure previously discussed to ensure equality.

head and the piston marked, align the piston in the combustion chamber of the head. From this point on, the procedure is the same as described above.

Fitting the piston dome

The real problem is what to do if the above process indicates that the piston will come in contact with the head. When I was working in a car dealership, one of the things I dreaded most was receiving a work order that called for the installation of roof racks and the like. These operations required holes to be drilled in brand new cars. If only the consumer really knew what I was doing to the second biggest investment of his or her life, he or she would certainly reconsider. There is always a hesitation to begin grinding on new cylinder heads and new pistons. Get over it!

There are a few things to remember in getting ready to grind away on those new parts that you just paid between $1,000 and $2,000 for. First, do not let your wife find the receipts; second, make sure she is at work or otherwise out of the house when you start grinding on what she is likely to interpret as her $2,000 diamond ring or sewing machine. With these critical items attended to, make sure that you have the proper tools for doing the grinding. A high-speed, high-horsepower grinder is necessary. As useful as the small Dremel motors are, they do not have enough horsepower to prevent the grinding bit from binding. If the grinding bit binds, it may gouge the head or piston, requiring more metal to be removed to repair the damage than you planned. Finally, if high compression is your goal, remember that every flake of metal removed from the piston or head lowers the compression.

Now, pull a large coin out of your pocket. You must make a decision about either removing

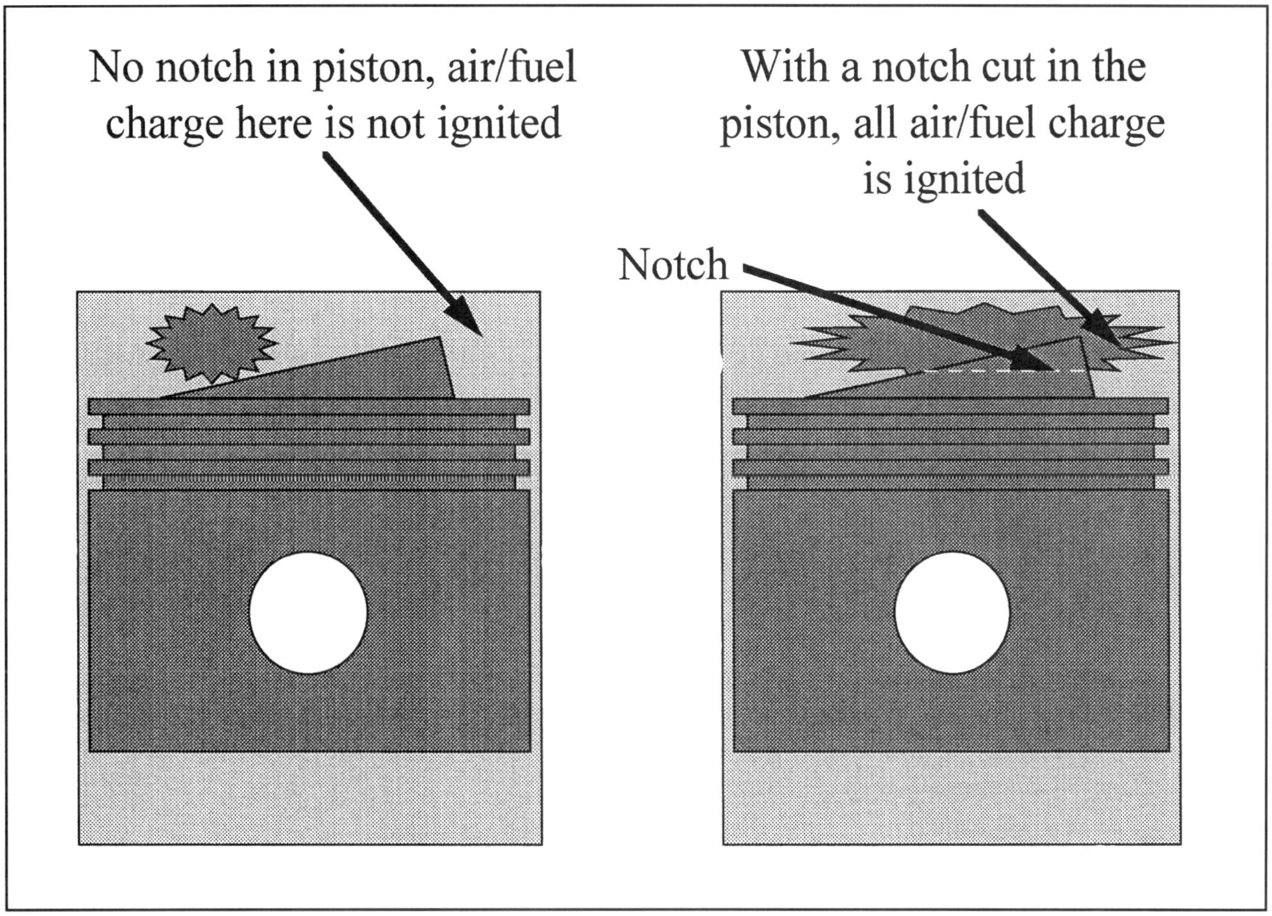

One of the major problems with the domed pistons is the ridge formed by the dome itself. This ridge prevents the flame front from traveling across the entire combustion chamber.

metal from the piston dome or from the combustion chamber. As the opening sentence of this paragraph implies, it makes little difference where the metal is removed. If you prefer some logical thought about it as opposed to the fall of a coin, considerations are the type and thickness of the head, and the size of the dome on the piston. Use the high-speed grinder to remove the least amount of metal necessary to provide clearance.

After grinding piston and combustion chambers of the head, it is time to waste another head gasket. Install the crankshaft using a set of good main bearings. Install the connecting rods on the pistons and install the piston/rod assemblies in the engine. Use a good but expendable set of rod bearings and torque the connecting rods. Before installing the heads, place some more modeling clay on the top of the piston. Be careful not to use an excessive amount of clay. This can result in a hydraulic lock that can damage the engine. (Note: If you have a friend who is a psychologist, do not allow him or her to witness this part of the operation; doubtless your friend would attach some Freudian meaning to the color of the clay, not to mention the application of the talcum powder to prevent sticking.) Now install the cylinder heads using the sacrificed set of head gaskets. Rotate the crankshaft through one complete engine cycle, two revolutions of the crankshaft. Sufficient clearance must take into account valve float, connecting rod stretch, and thermal expansion of all components. The general rule is there must be at least 0.080in of clearance for the intake valves and 0.100in for the exhaust valves.

Repeat the grinding process until proper clearance is achieved. Remember that it is easy to remove metal, but it is very difficult to put it back once removed. Wasting an extra set of head gaskets as a result of being conservative on the removal of metal from the pistons and cylinder head is a lot cheaper than replacing the heads or pistons.

Relieving the piston dome

Many years ago, I took a few flying lessons in the town of Renton, Washington. Had I known

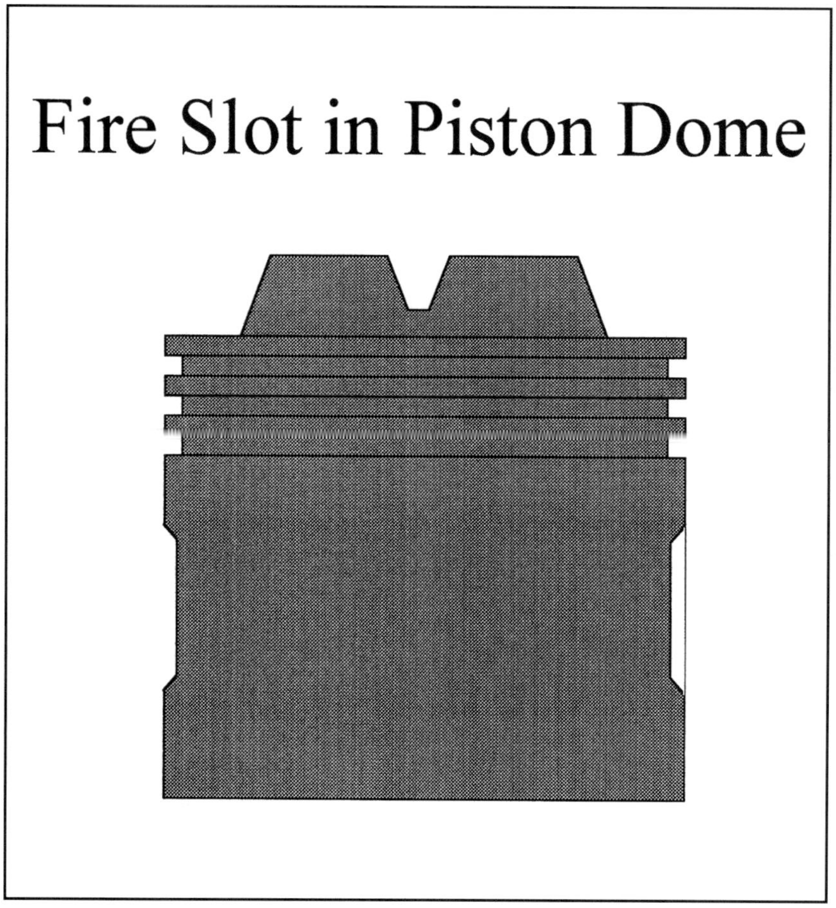

Cutting a fire slot, or tunnel, in the piston dome provides a path for the flame front to travel to the portion of the combustion chamber that is masked from the spark plug.

then what I know now about this particular airfield, I would have probably found another place to take the lessons. It seems that this was the last airfield visited by Wiley Post and Will Rogers before their fatal trip to Alaska. During the course of the lesson, the instructor had me fly toward a mountain. The idea was to get me used to the fact that winds aloft may force you to steer to the left or right of an object in order to actually fly toward the object. Having been raised in an Air Force family I found this to be a "Well, duh!" exercise. What I did not like about the exercise is the fact that the instructor was having me approach the mountain at what I felt was an insufficient altitude. However, I noticed that the mountain had a notch or saddle at the midpoint of the crest. I aimed for the midpoint.

Imagine the little Cessna I was flying to be the flame front. There is an air/fuel charge on both sides of the piston dome, yet the piston dome is much like the crest line of that mountain without the saddle. Imagine also that your license only permits you to fly VFR (visual flight rules) and the cloud cover is almost touching the mountain like the cylinder head almost touches the dome of the piston. How is the flame front to get from the spark plug side of the piston dome to the other side of the piston dome? At the point of ignition, the piston is at top dead center and the dome of the piston is dividing the combustion chamber.

Although a rare requirement, it is sometimes necessary to cut a "fire slot" in the piston dome. In pistons with extremely high domes, the dome of the piston may mask the spark plug from the rest of the combustion chamber. This masking will at least contribute to poor combustion and may prevent the engine from running. After the famous Mustang was built, we decided to do some testing. We went to the local drag strip. At about 3000rpm the engine would begin to misfire. This problem continued throughout our use of this engine. Only years later did I find out that the flame front was probably masked, kept from traveling the total diameter of the combustion chamber by the dome of the piston. Therefore, one part of the air/fuel charge was burned, and the engine ran out of power at a relatively low rpm.

The procedure for cutting a fire slot, or tunnel, in the piston dome is like the procedure for the removal of any metal from an expensive component. Begin with a sacrifice to Synchro the god of all things needlessly damaged and therefore useless. Using a high-speed electric grinding motor or an air-powered grinding motor, grind a notch in the dome which will allow the spark across the gap to be open and exposed to the entire combustion chamber when the piston is at top dead center.

Checking piston thickness

There are few things that can happen to a piston that are as potentially catastrophic as detonation. Detonation occurs when high temperatures in the combustion chamber are ignited by the heat of compression. This causes the gases from combustion to begin to expand as the piston is traveling toward the top of the cylinder. This means that the piston must now compress an expanding gas. The result is the violent collision of an expanding gas and metal traveling at several thousand feet per minute.

If the piston head is too thin, it will not be able to withstand the extreme pressures of detonation, resulting in catastrophic failure of the piston, metal fragments flying around inside the block, and local auto parts store cash registers drooling at the opportunity to sell you a large number of engine parts. To measure the thickness of the piston head, you could place the piston head down on the surface of the Moon. Call a buddy at NASA and have him point a laser at the underside of the piston head. Divide the time it takes for the light to return to the site by two, then divide that quotient by the speed of light. That will give you the distance from the NASA facility to the bottom of the piston head. Now move the piston and repeat the procedure. This will be the distance to the surface of the Moon. Subtract the distance from the NASA facility to the bottom of the piston from the distance from the NASA facility to the surface of the Moon. This will give you the piston head thickness. If you do not know anyone at NASA, there is an alternative procedure.

Mount a long center punch or similar pointed object in a vise. This will form the lower anvil for measuring the thickness of the head of the piston. Place a dial indicator on a holder so that the piston or any other object can be measured between. Adjust the dial indicator so that it reads zero against the point of the center punch. Place the piston in the measuring tool you have just made. Measure several of what you feel are the thinnest points in the head of the piston. The thinnest point on the head of the piston should not be less than 0.080in if the engine is to be run for less than 15 seconds at a time down a track only one fourth of a mile long. If you are going to be running any longer than 15 seconds, the piston head must be thicker.

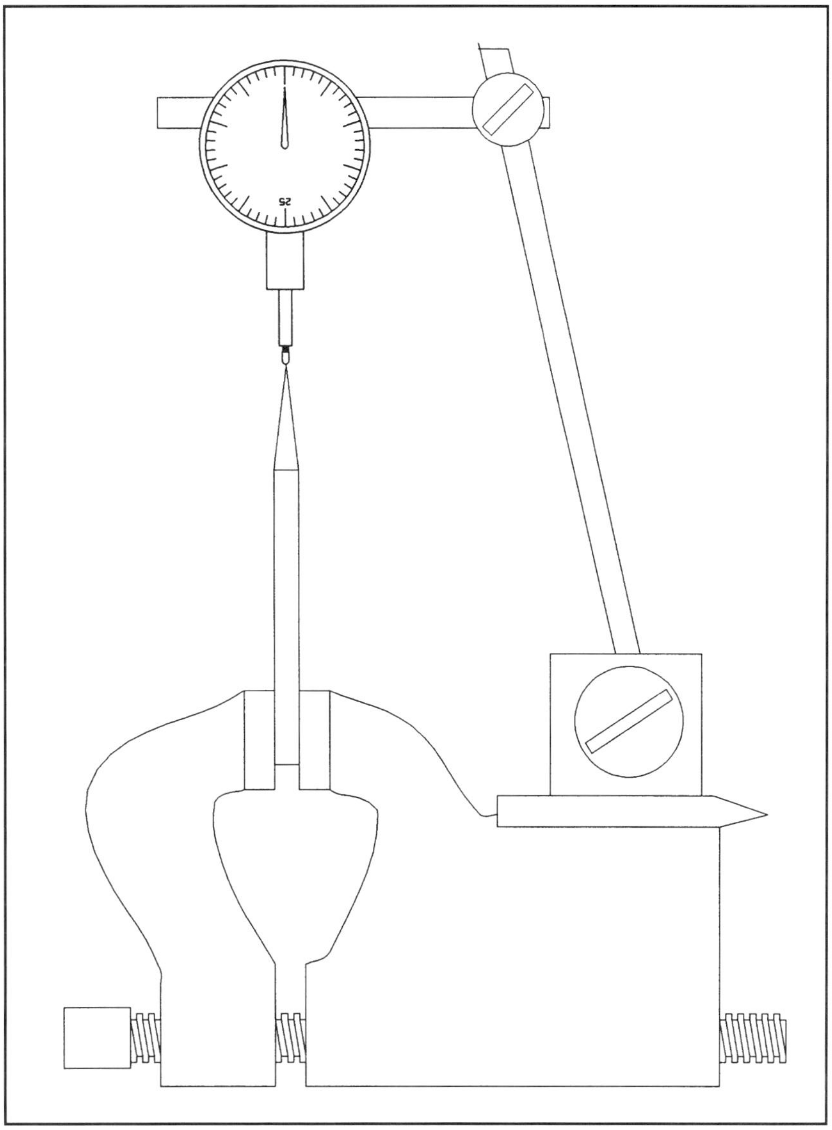

A vise, a dial indicator, and a long pointed punch can make an excellent tool to check piston thickness. This measurement is important to guard against casting mistakes on new pistons. This measurement is even more important when the piston has been modified through grinding.

A more modern approach to measuring the piston head thickness is the use of a sonic measuring tool. These devices use sound waves to measure the thickness of metal. Check with some of the machine shops around town. They probably have one and can provide you with accurate measurements.

There are no hard and fast rules about piston head thickness, and you will really only need to check it when planning to modify the head of the piston and when you have finished modifying the head of a piston. I have two pieces of best advice. First, do not modify the head of the piston so that it is thinner at any point than it was at the thinnest point before you began. Second, call the manufacturer of the piston, explain

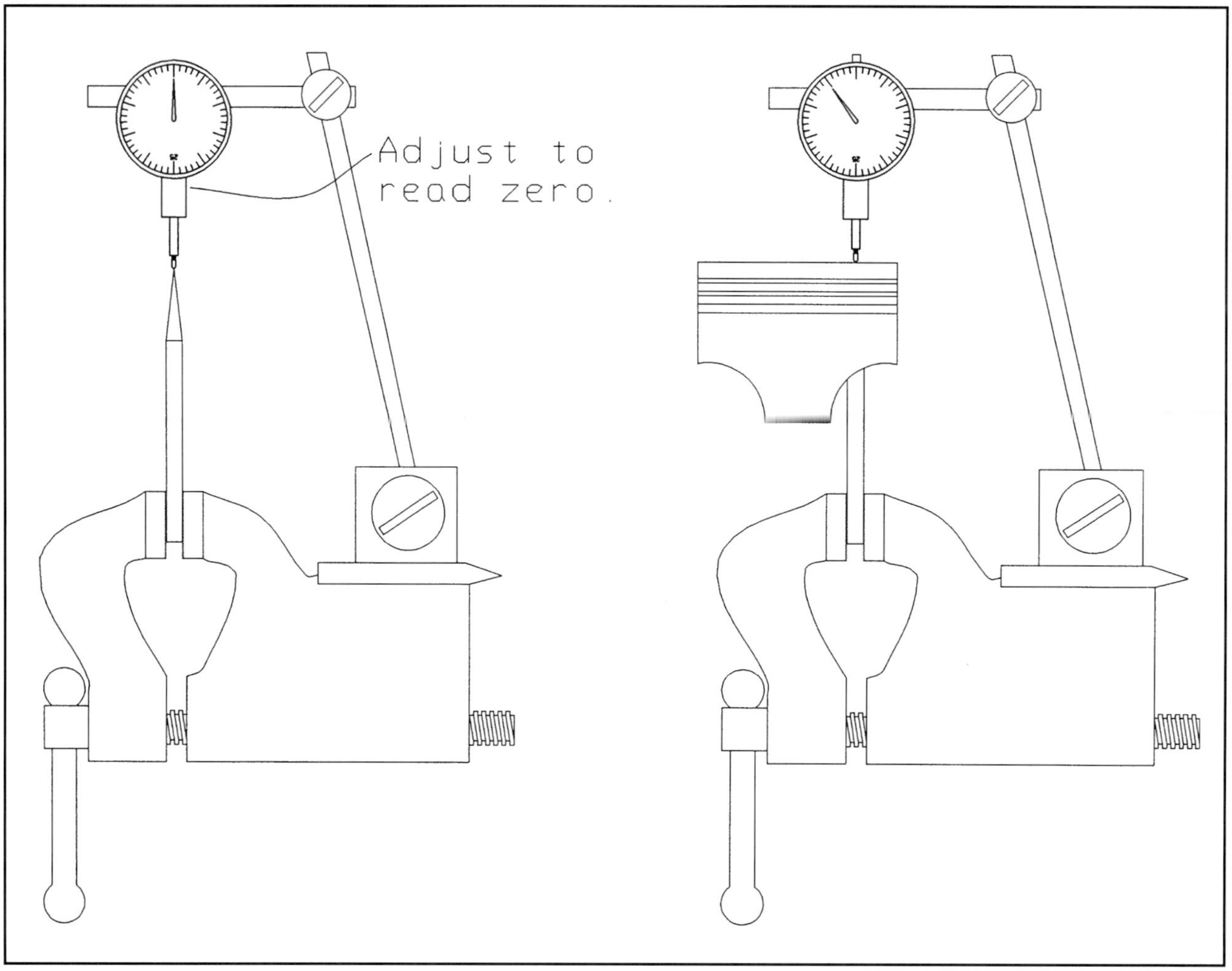

Adjust the dial indicator so that it reads 0 against the point of the center punch. Place the piston in the measuring tool you have just made. Measure several of what you feel are the thinnest points in the head of the piston. The thinnest point on the head of the piston should not be less than 0.080 if the engine is to be run for less than 15 seconds at a time down a track only one-quarter-mile long. If you are going to be running any longer, the piston head must be thicker.

the type of use for the engine you are building, and ask the manufacturer for its minimum recommended thickness. The manufacturer will take into consideration the type of fuel you are using, the metallurgy of the piston, the speed of the piston at maximum rpm, mean and maximum cylinder pressure, how long the engine will run at one time, and the typical and maximum loads on the engine. They will calculate these factors on a Cray Supercomputer and then . . . guess. Remember this is an area where the faint of heart will save money in the long run.

Most readers of this book are probably not going to be doing any serious piston modifications. Therefore, the typical reader might feel that this is an operation they can skip. After all, they might say, I just paid $950 for this set of “Agamemnon Superforged” pistons. They cannot possibly have a defect like a thin casting. You might find this a rather depressing presumption when you are sitting on the side of the road or track with fragments of your $950 pistons lying in the bottom of the oil pan.

Piston lightening

In most cases, the only reason to lighten the piston is to balance the piston. Find a scale accurate to a gram. Find the lightest piston of the group you are using. Remove metal from the heavier pistons until they weigh the same as the lightest. Remove the metal from the pistons according to the guidelines described below. For most enthusiasts, the best way to ensure a properly balanced set of pistons is to pay a few bucks extra to the manufacturer. Manufacturers take the time to weigh their pistons and match them in sets by weight.

The ideal piston would be extremely light and extremely strong. There has been a great deal of experimenting over the years with ceramics as a way of meeting this goal. The lighter the piston, the lesser the inertial forces acting on the piston, but the weaker the piston.

First of all, lightening the pistons only increases throttle response. It has no place when the pistons are to be used in an engine destined for a vehicle running up and down Alaska's Dalton Highway (known locally as the "haul road"). The Dalton Highway is nearly 500 miles of gravel heading north out of Fairbanks to Prudhoe Bay on the Arctic Ocean. There are two places on this stretch of road to buy supplies and gasoline. Lightening the pistons inherently makes them weaker and more fragile. This is no place for an engine with lightened pistons.

On the other hand, when the ability for the engine to change speeds suddenly and rapidly is more critical than the ability to hang together for a hundred thousand miles, there is a lot that can be done to lighten most pistons. If you took the time and spent the money to buy quality performance pistons, the only way to significantly and safely reduce the weight is with the knowledge and trial-and-error experience of a skilled performance machinist. The machinist you would want to use will be a machinist that has a secret logbook of piston lightening successes and failures. In this skill, near success is when you removed one gram too much from the piston and it came apart. Total success is when you build a new engine using the same type of piston and do not remove that critical last gram. These types of successes and failures are what makes that machinist the one you want to use.

If you are one who is inclined toward do-it-yourself without the

Even though the manufacturer of these pistons never planned for them to be used in a performance engine, rest assured that some effort was made to ensure consistent weight. For maximum performance, they should be balanced to within 1/2 gram of one another. Remember that performance does not only mean how fast it can get down the track, it also refers to fuel economy and smoothness.

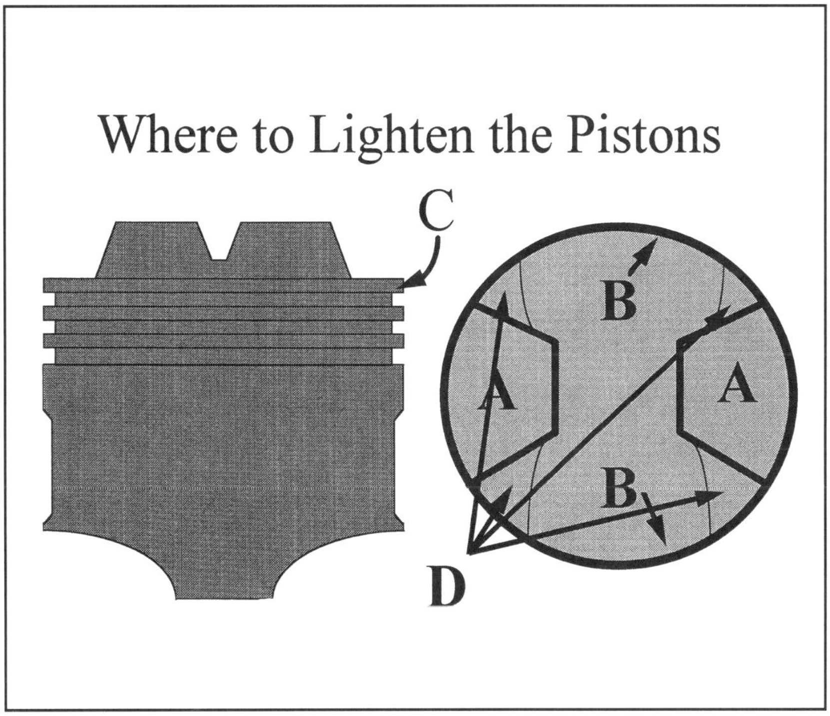

There are four good places to lighten pistons: (a) the wrist pin bosses; (b) the inside of the piston skirt; (c) the underside of the ring land; and (d) the piston struts next to the wrist pin bosses.

aforementioned experience, there are four places where weight is typically removed:

1. Remove metal from the inside of the skirt.
2. Remove metal from the underside of the piston head near the wrist pin bosses.
3. Remove metal from the underside of the ring lands on the pistons.
4. Remove metal on the piston struts next to the wrist pin bosses.

Two things are critical to this operation: Use radiused cutters. Radiused cutters feature rounded edges on the cutting tip. The rounded edges prevent sharp-edged gouges from occurring as the metal is being removed. Extreme caution should be taken to leave no sharp edges! Sharp edges are the places where stress

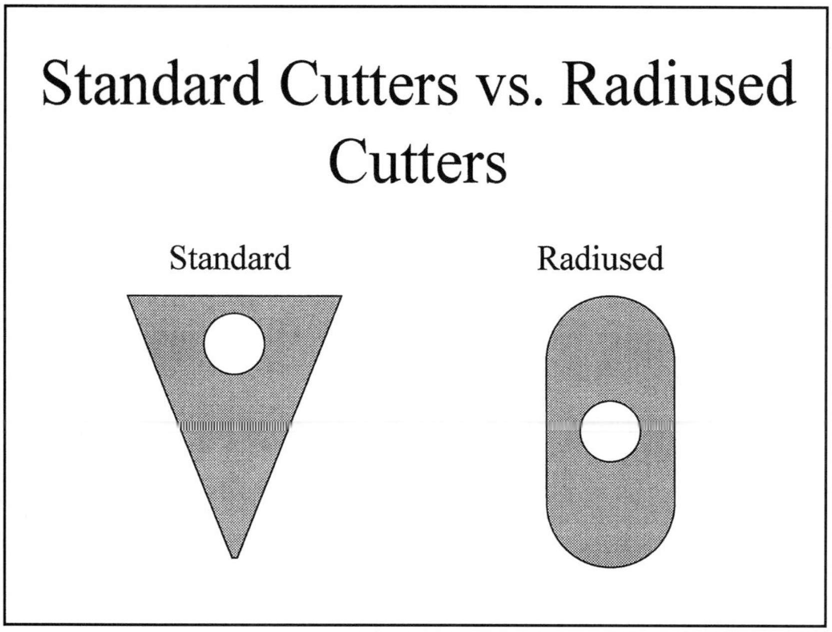

Use radiused cutters. Radiused cutters feature rounded edges on the cutting tip. The rounded edges prevent sharp-edged gouges from occurring as the metal is being removed. Extreme caution should be taken to leave no sharp edges. Sharp edges are the places where stress fractures and cracks can start.

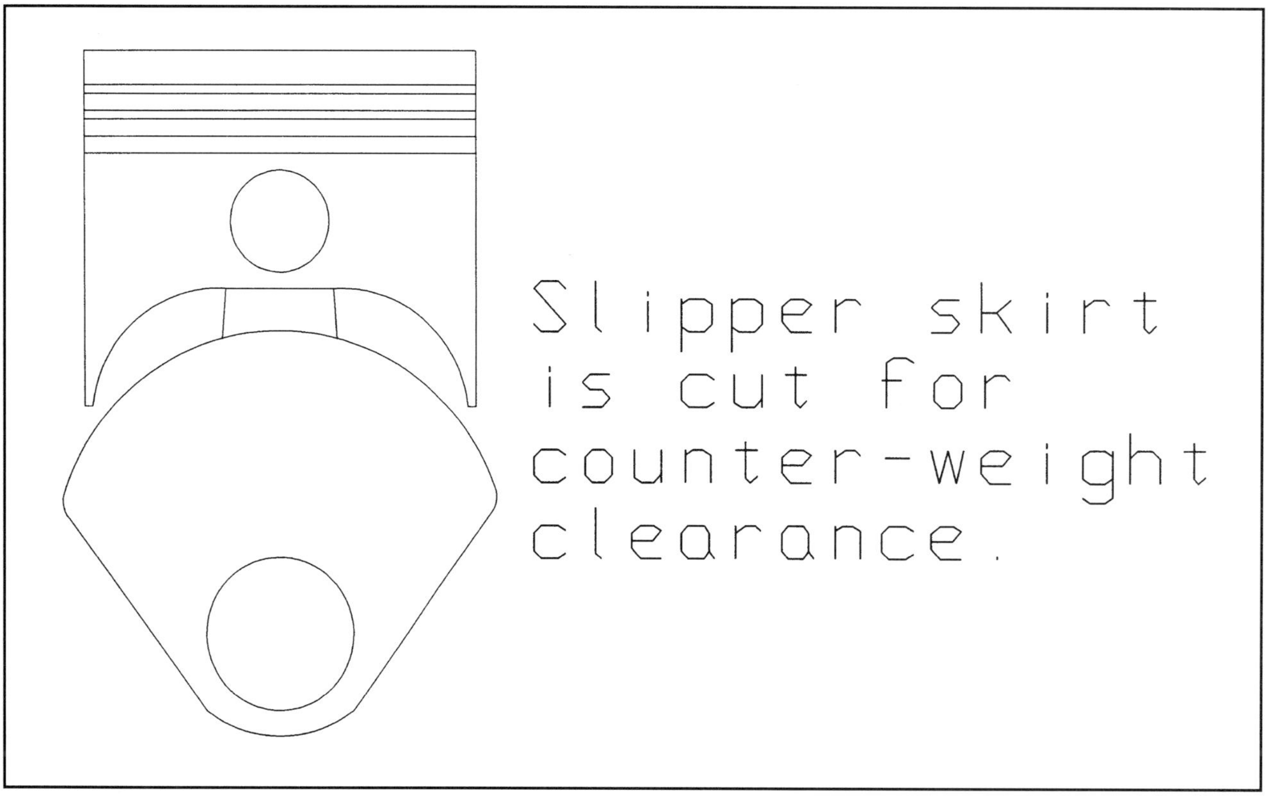

A critical clearance that is easy to miss is the piston skirt-to-crankshaft clearance.

fractures and cracks can start.

A final benefit of the lightening process is not only does it reduce the reciprocating mass, it also allows lightening of crankshaft counterweights, which reduces the rotating mass. The reduction in rotating mass allows for more rapid changes in rotational speed. Bottom line: lower mass equates immediately to greater changes in speed.

Skirt-to-crankshaft clearance

The slipper skirt piston was developed when the crankshafts began to grow counterweights. The skirt of the standard trunk (full skirt) piston tried to occupy the same space as the crankshaft counterweights.

Gas porting

Gas porting the piston involves precision drilling tiny holes in the top of the piston so that the gases that would otherwise be trapped between the top compression ring and the head of the piston can escape. Located at the back of the ring groove, these holes are typically 0.040 to 0.060in in diameter. The holes must be evenly spaced. Only pistons destined for running a quarter mile at a time should be gas ported. In these engine rings, side clearance is often less than 0.002in. The gas ports are needed to prevent tremendous pressures from building under the top ring land and tearing the head off the piston at extreme rpm.

Choosing performance pistons

The first question you need to ask yourself is how fast the

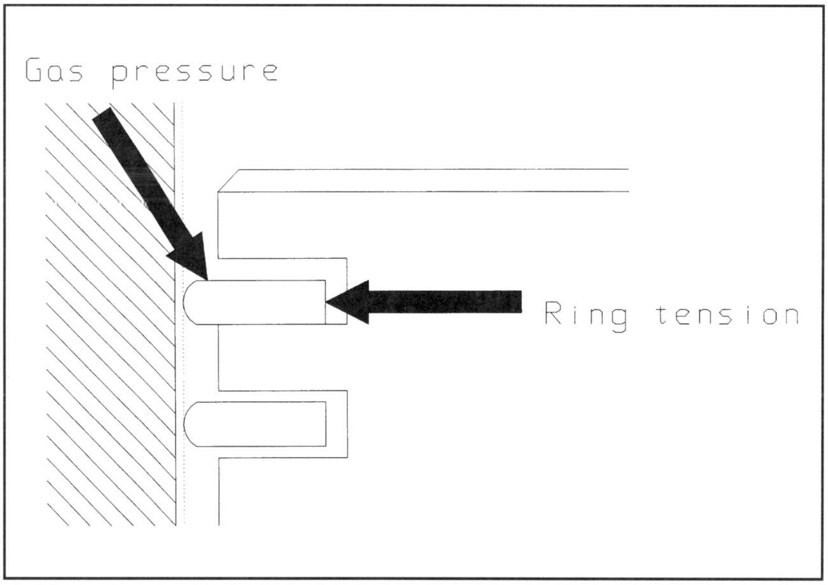

As the engine runs, especially as it runs at high speeds under high loads, gases can become trapped between the top compression ring and the piston head. These trapped gases can do severe damage to the piston. Gas porting is the solution to this problem.

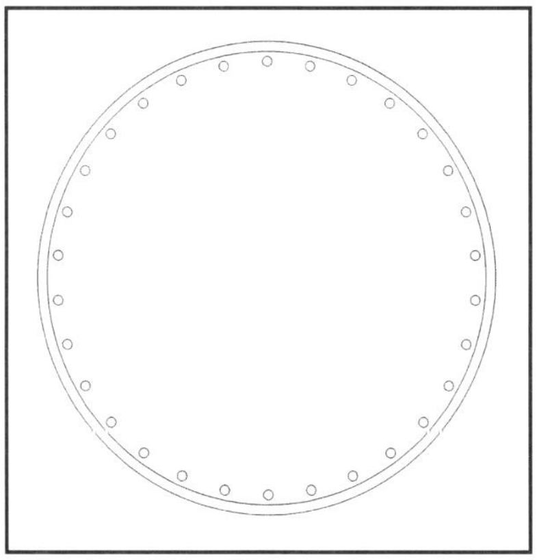

Located at the back of the ring groove, these holes are typically 0.040 to 0.060in in diameter. The holes must be evenly spaced. Only pistons destined for running a quarter mile at a time should be gas ported. In these engines, ring side clearance is often less than 0.002in.

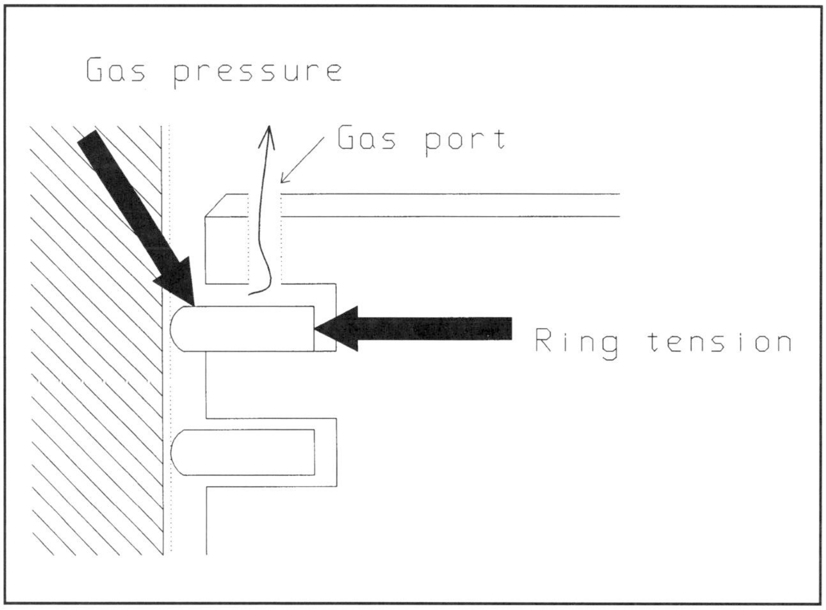

Gas porting the piston involves precision drilling tiny holes in the top of the piston so that the gases that would otherwise be trapped between the top compression ring and the head of the piston can escape.

engine will be running. A drag fueler may run 10,000 rpm. The pistons you purchase for such a rig must be able to hold together at that speed.

Expansion and cold piston clearance

As the piston warms during operation, it expands. Most pistons are designed with a "cam" ground design. This means that the diameter of the piston in one direction is slightly smaller than the diameter of the piston in the other direction. The design of the piston, the mass concentrations of the piston, are such that the expansion rates are different on the two axes. When the piston is at operating temperature, the differences in expansion have caused the two axes to become equal in diameter. The piston is now round. This presents us with a problem, however, when we are trying to measure the piston clearance. The measurement for piston clearance will, out of necessity, be taken cold. Therefore, the clearance specs may seem very large to the layman. It would be a mistake to think that power could be improved by tightening up this specification. With a tight piston-to-cylinder clearance, the expansion of the piston as it warms could cause the piston to tend to seize in the cylinder.

Wrist pins (piston pin)

The wrist pins are in no way exempt from the stresses and strains of the reciprocating piston and rod assembly. The job of the wrist pins is to transfer the energy from the piston to the connecting rod. The wrist pin is constructed from high-grade steel. For the most part, you need to worry very little about the choice of wrist pins. When either stock or performance pistons are purchased, they come with wrist pins adequate to handle the loads projected by the manufacturer of that piston. This load is typically not more than 8,000psi. The design of a stock piston permits the assembly to withstand this kind of punishment for only a short time. Many racing pistons are designed to allow the piston-to-pin contact surface to handle as much as 20,000psi.

Let us do a little calculation concerning combustion loads. Let us consider the piston to be the moving floor of a cylinder. Let us imagine a sealed container with no moving floor. The pressure potential in the cylinder is about 1,743psi after combustion. This assumes that the pressure in the cylinder was a normal atmospheric pressure of 14.7psi prior to the initial precombustion pressurization. If the diameter of the cylinder were 4in and if that load were to be transferred to the 2sq-in contact area of the wrist pin, that wrist pin would be subjected to a force of nearly 44,000psi. Note the chart below.

But we are dealing with a chamber that has a moving floor. The pressures in the cylinder will not build as high as described above because as the pressure begins to build the floor below, the piston will begin to drop away. As a result, the pressure we are concerned with is the mean effective pressure of the cylinder. MEP, or mean effective pressure, is calculated by multiplying the horsepower of the engine by 792,000, then dividing by the displacement of the engine times the rpm of the engine. If you are curious about the 792,000, it is not a number pulled out of the air. It is a constant based on the 12.1 conversion factor of inches to feet, times the two crankshaft rotations it takes to complete a full set of cylinder cycles on a four-cycle engine, times 33,000, the number of pounds-feet per minute equal to one horsepower.

$$\text{mep} = \frac{\text{hp} \times 792{,}000}{\text{displacement} \times \text{rpm}}$$

As you can see from the numbers, the forces exerted on the wrist pin are relatively light and do not increase dramatically as either the horsepower or the rpm increase. Note that when the horsepower increases from a relatively low horsepower to a horsepower eight times greater, the pressure on the piston wrist pin is only increased fivefold. At 2,000hp and an rpm of 10,000, the pressure on the wrist pin is barely out of the acceptable range for a stock unit. This means that the highly resilient pins, those that come with performance pistons, can be shortened. This reduces the contact area and therefore increases the

Charging pressure	Compression ratio	Compression pressure (psi)	Combustion pressure (psi)	Piston surface area	To wrist pin load moment
14.7	8	243.49	905.06	50.24	22735.20
14.7	9	285.46	1061.05	50.24	26653.53
14.7	10	329.09	1223.23	50.24	30727.51
14.7	11	374.28	1391.19	50.24	34946.80
14.7	12	420.93	1564.60	50.24	39302.67
14.7	13	468.97	1743.14	50.24	43787.57

psi on the pin; however, it can significantly decrease the reciprocating mass. How important is this? Is it worth the fact that it will probably void the warranty on the piston/pin assembly? Frankly, unless you are trying to squeeze the last hundredth of a second out of your drag car on the quarter mile, it is hardly worth the effort or the loss of the warranty. However, in the 2,000hp example above, a wrist pin designed to handle 20,000psi can easily be trimmed of half its contact area with the piston and still maintain a margin of safety.

The wrist pin is relatively trouble free in spite of the repeated forces exerted on it. Many years ago when I owned a repair shop in Bellevue, Washington, I arranged a car purchase between one of my employees, Pat, and one of my customers. The car's engine had a tapping noise, which is why the customer wanted to get rid of it. Soon the tapping noise became a rapping noise. Pat turned up his radio. A short time after that the rapping noise became a knocking noise. Pat turned up his radio still louder. After several months, Pat had the choice of either buying bigger and louder speakers for his radio, or rebuilding his engine. Against my advice, Pat chose to rebuild the engine. When the engine was disassembled, we found that a retaining clip for the wrist pin had broken at some point in the distant past. The wrist pin had slipped off to one side and had slowly routed a deep groove in the cylinder wall. The block was useless and had to be replaced. Pat told me he was sorry he had not heeded my advice and bought the new speakers.

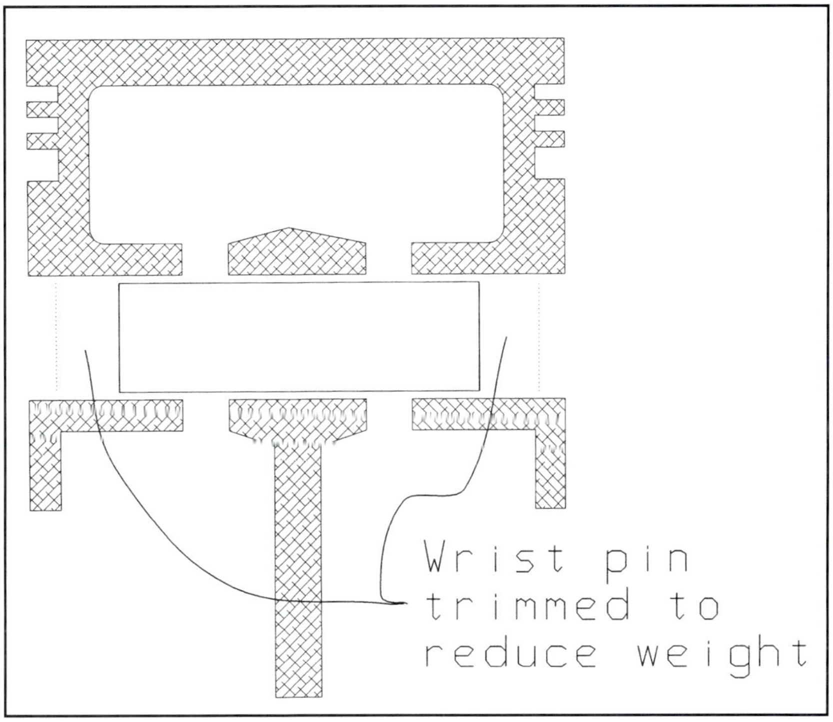

The typical piston pin, or wrist pin, is capable of withstanding much greater dynamic forces than that to which they are normally subjected. This means that in engines where the ability to accelerate is of primary interest, reciprocating mass can be reduced by shortening the wrist pins.

Full-floating

Full-floating piston pins are held into the piston with lock rings. These lock rings might be simple wire rings. These are the cheapest and least dependable. Higher quality lock rings are used in most pistons. Since the pins are not press fitted into either the piston or into the connecting rod, the survival of the engine depends on these little clips. Failure of the clips dooms the engine to at least the aforementioned "Pat" syndrome.

Installing the pistons onto the rods in the full-floating wrist pin design is easy, since the wrist pin slides through both the piston and rod with pressure applied by even the frailest of thumbs.

Semifloating

Rod fitted

An alternative method of holding the wrist pin in place is to use a press fit into the small

Horsepower	Displacement	RPM	Mean effective pressure	Rod pressure
250	350	6000	94.29	2357.14
300	350	6000	113.14	2828.57
350	350	6000	132.00	3300.00
400	350	6000	150.86	3771.43
450	350	6000	169.71	4242.86
500	350	6000	188.57	4714.29
550	350	6000	207.43	5185.71
600	350	6000	226.29	5657.14
650	350	6000	245.14	6128.57
700	350	6000	264.00	6600.00
2000	455	110000	348.13	8703.30

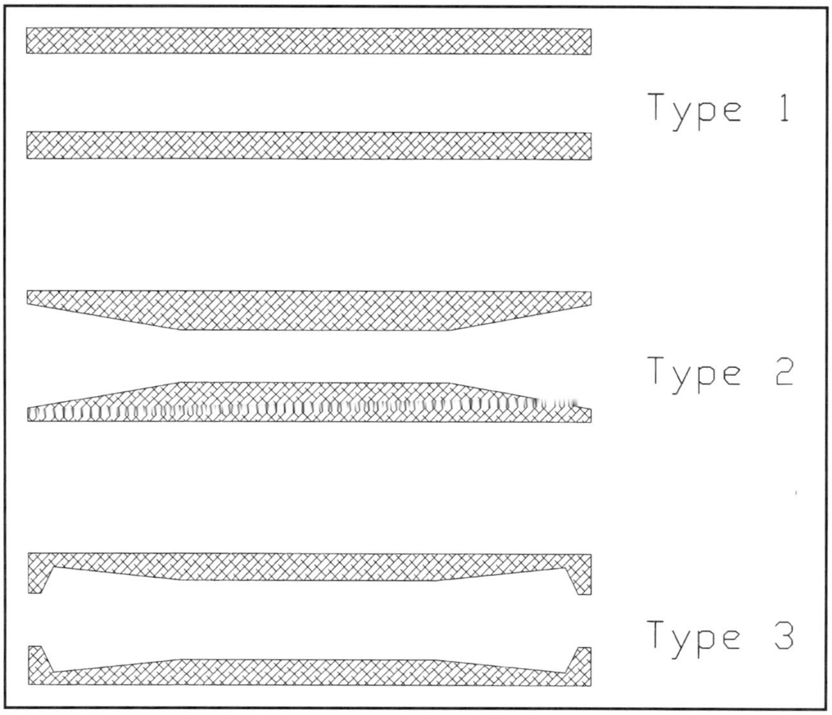

Some performance piston manufacturers will alter the design of the basic wrist pin to reduce weight. This does not preclude further shortening.

end of the connecting rod. The position of the wrist pin, therefore, remains stationary relative to the connecting rod and allows the piston to float. The pin is pressed into the small end of the rod with an interference fit of 0.001in. An interference fit means that the object to be inserted into the connecting rod, in this case the wrist pin, is larger than the hole in the small end of the connecting rod. This interference fit means that the wrist must be pressed into position with more than two tons of force.

This semifloating wrist pin is more of a problem to remove and insert than the full-floating wrist pin. The safest route is to have the machine shop install semifloating wrist pins. A press tool can be made to facilitate the pressing of wrist pins in semifloating applications. Using a piece of hardwood, carve a cradle

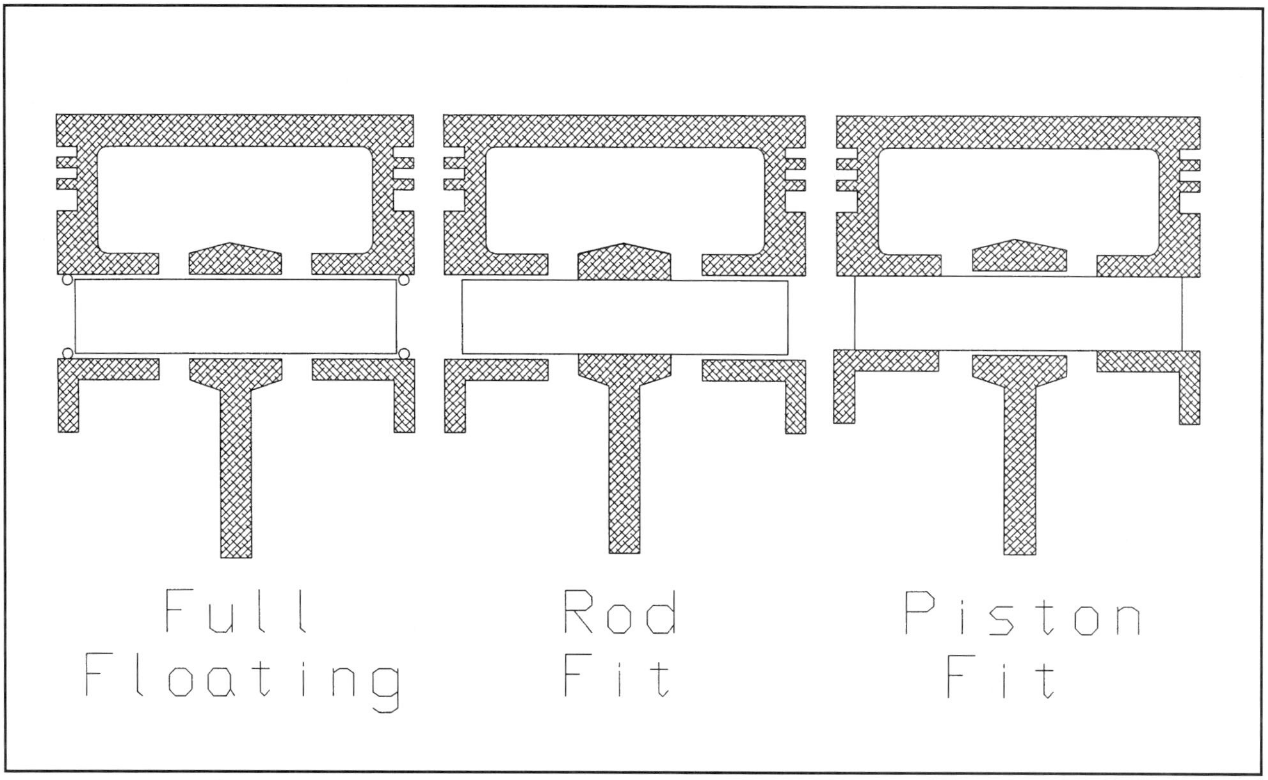

There are three ways that the wrist pin is fitted into the connecting rod/piston assembly. Full-floating is where the wrist pin is not pressed into either the connecting rod or the piston. It rides on a film of oil in each. In a rod-fitted assembly, the wrist pin is press fitted into the connecting rod and floats in the piston. Piston fit presses the pin into the piston, allowing the rod to float free.

the diameter of the piston. This will support the piston evenly as the wrist pin is pressed into place. These cradles can also be purchased commercially. Of course, allowing the machinist to install the wrist pins on semi-floating applications may be the best answer.

Piston fitted

There are a few applications that feature a wrist pin that is

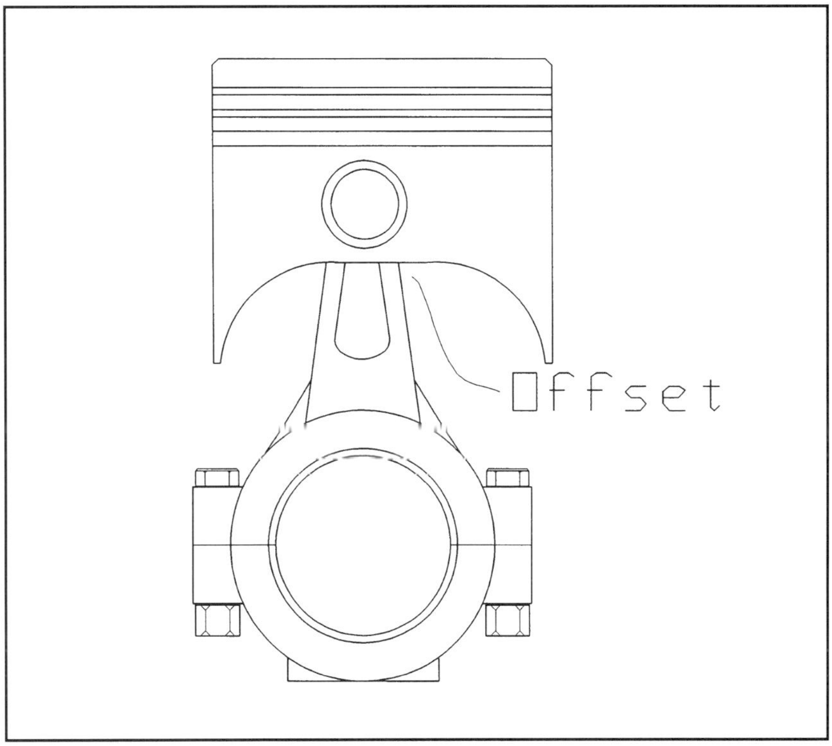

Right
Remember that pistons and connecting rods are often asymmetrical. Be careful when assembling these units to ensure they are assembled properly with respect to one another. Improper assembly can result in the improper side of the piston being asked to act as the thrust surface. The connecting rod or piston skirt may also tap other internal components, leading to potential for catastrophic failure.

A full-floating wrist pin design can be assembled using the thumb as a press.

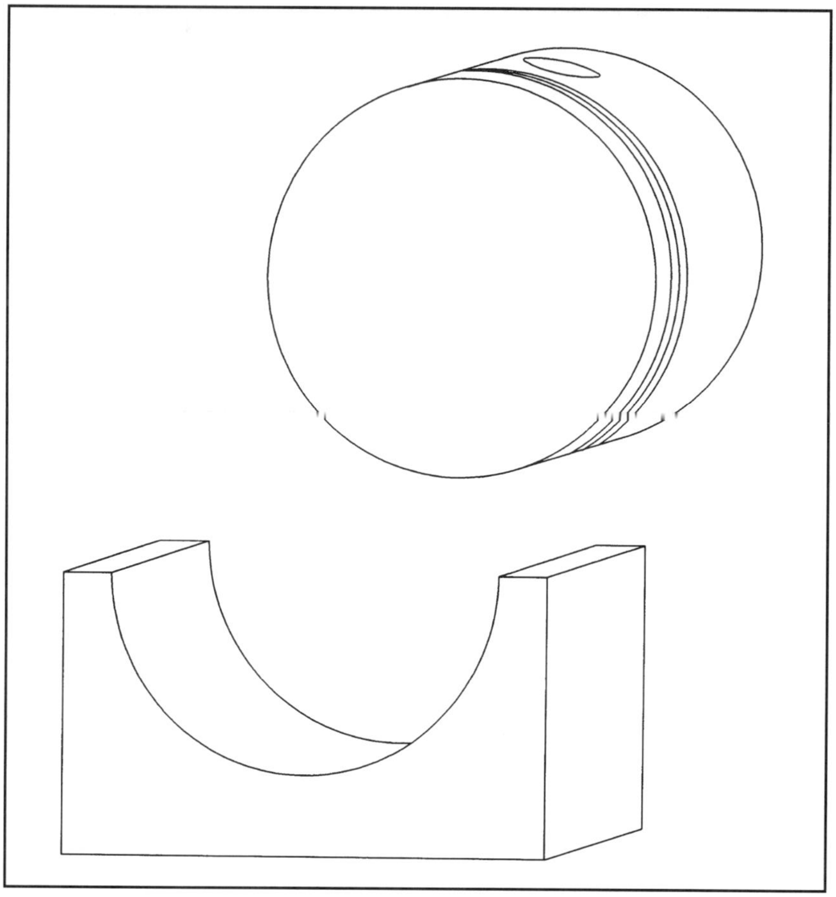

If the wrist pin design is either piston pressed or connecting rod pressed, a cradle and hydraulic press may need to be used for assembly.

pressed into the piston but floats free on the connecting rod's small end.

Another thing to remember

Pistons are directional. In 1978, I had the fortune, or perhaps misfortune, of rebuilding a Volvo B-28 engine. This engine is particularly interesting because when fully disassembled, you are left with a workbench full of small pieces. This engine had a bent connecting rod. It had been bent during a rainstorm. This happened in Fort Worth, Texas. You are probably aware that Seattle has a reputation for a lot of rain. Fort Worth, however, raises images of the semi-arid or even of the desert. Several times a year, the citizens of Fort Worth are blessed with torrential rains. The owner of this particular Volvo lived in a low area. After one of the more torrential rains of the 1970s, he awoke to find the car almost completely submerged. Later that day he discovered that the engine would not start. The battery was dead, or at least the engine would not turn over. He towed the car to the top of a hill. Turning on the ignition switch and placing the transmission in second gear, he allowed the car to roll down the hill gathering speed. When the car reached about 35mph, he popped the clutch. The engine did not start. He repeated this process several times. Finally he gave up and had the car brought in for a "no-start" problem. Unknown to the car owner, the number one cylinder was full of water and near top dead center on the compression stroke. When he popped the clutch, the connecting rod bent.

After rebuilding the engine, I started it. There was a loud knocking noise. When I disassembled the engine again I found that I had installed the number one connecting rod on the number one piston backwards. Every time the piston went to bottom dead center, the piston would touch or tap the top edge of the connecting rod, producing the noise. Pistons and connecting rods have direction.

Wrist pin oiling

The wrist pin is probably one of the most difficult of the high-stress components of the engine to lubricate. Moving up and down the cylinder 500 to 5,000 times per minute or more, it does not stand still to be well lubricated.

On stock pistons, the wrist pin bore is smooth with a hole drilled in the bottom of the bore toward the crankshaft. As the crankshaft rotates, oil is splashed toward the piston. The oil is then expected to migrate along the wrist pin to form the lubricating film between the pin and the piston. This actually works pretty good on a low-performance stock engine. However, if you have in mind racing your newly rebuilt engine, this type of oiling might not be good enough.

A second method takes oil picked up by the oil scraper ring and forces it into a groove cut in the piston pin bore. This groove helps move the oil along the length of the wrist pin. On the planet Palindrome, located in the Gamma quadrant of the Alpha hemisphere of the Milky Way Galaxy, this oiling method for the wrist pin is totally superior. On that planet combustion is always perfect. On planet Earth combustion is not always perfect. Partially burned hydrocarbon particles can be picked up by the oil scraper ring and carried along with the wrist pin lubricating oil into the groove. After a few thousand miles, the groove becomes restricted and will eventually cut off wrist pin lubrication.

No matter the type of wrist pin lubrication system your old or new pistons were designed with, almost all of them can be improved. Drill a hole perpendicular to the flat part of the wrist pin boss on the underside of the piston. The hole should be 0.125in and drilled so that it will intersect the wrist pin bore at its centerline.

Wrist pin and bore location

When the connecting rod is pushing the piston up the cylinder, it does so at an angle. This causes the piston to tilt, pushing the skirt of the piston against the cylinder wall. When the spark plug fires the air/fuel charge, the piston is pushed down the cylinder, forcing the connecting rod and crankshaft to move. The angle of the connecting rod now causes the piston to rock in the opposite direction. The closer the wrist pin is to the top of the piston, the less rocking will occur. The closer the wrist pin is to the top of the piston, the longer the connecting rod has to be. The longer the connecting rod is, the less sharp the angle at which it applies lateral force to the piston is, and therefore, less tilting of the piston will occur. Tilting increases the friction between the piston and the cylinder walls. This reduces potential engine power and increases the stress and strain on the piston and cylinder.

If the wrist pin were located on the same plane as the highest part of the piston, actually con tacting the cylinder wall, rocking of the piston on the compression stroke would be minimized. However, pressure on the top edge of the piston contacting the cylinder wall will be maximized. Again, a compromise is required.

Wrist pin bore

There is an old "Star Trek" episode called "Amok Time"

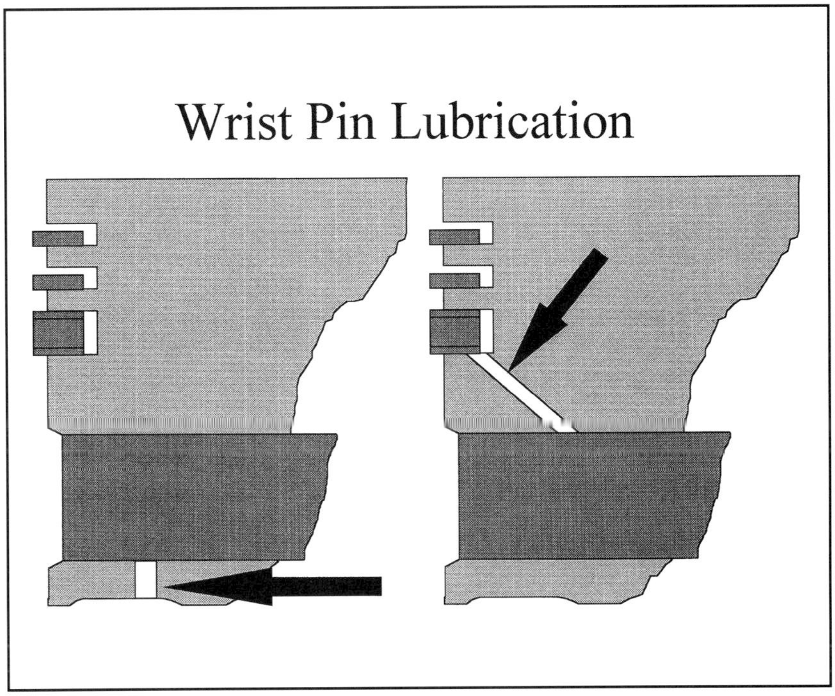

The wrist pin bore is smooth on stock pistons with a hole drilled in the bottom of the bore toward the crankshaft. As the crankshaft rotates, oil is splashed toward the piston. The oil is then expected to migrate along the wrist pin to form the lubricating film between the pin and the piston. This actually works pretty good on a low-performance stock engine. However, if you have in mind racing your newly rebuilt engine, this type of oiling might not be good enough.

A second method forces oil picked up by the oil scraper ring into a groove cut in the piston pin bore. This groove helps move the oil along the length of the wrist pin. However, partially burned hydrocarbon particles can be picked up by the oil scraper ring and carried along with the wrist pin lubricating oil into the groove. After a few thousand miles, the groove becomes restricted and will eventually cut off wrist pin lubrication.

where Spock gets all lusty for his betrothed. Kirk decides to protect Spock in a battle for the affections of the lovely Vulcan in question by fighting him to the death. Since the air on Vulcan is hot and thin, Kirk soon has difficulty supplying his lungs with enough oxygen to properly convert the food he had eaten into the energy necessary to kill Spock. Dr. McCoy complains about the unfair advantage Spock has in being more accustomed to the Vulcan atmospheric pressure and climate. The Vulcan leader, T'pau, responds, "The air is the air; what is to be done?" The location of the wrist pin bore is the location of the wrist pin bore, what is to be done? Very little is to be done. If you have plenty of money, money enough to afford failure, the wrist pins can be set closer to the head of the piston. This may improve the thrust angle. Bore the wrist pin bore off-center toward the head of the piston for an oversized wrist pin. This will locate the wrist pin closer to the head of the piston by 50 percent of the increase on the wrist pin diameter. All of this work may give you a better thrust angle, but at a price that will be paid in the loss of compression ratio. More work will now need to be done to boost the compression ratio back up to ideal.

A lot of work or expense for only a little improvement. Your

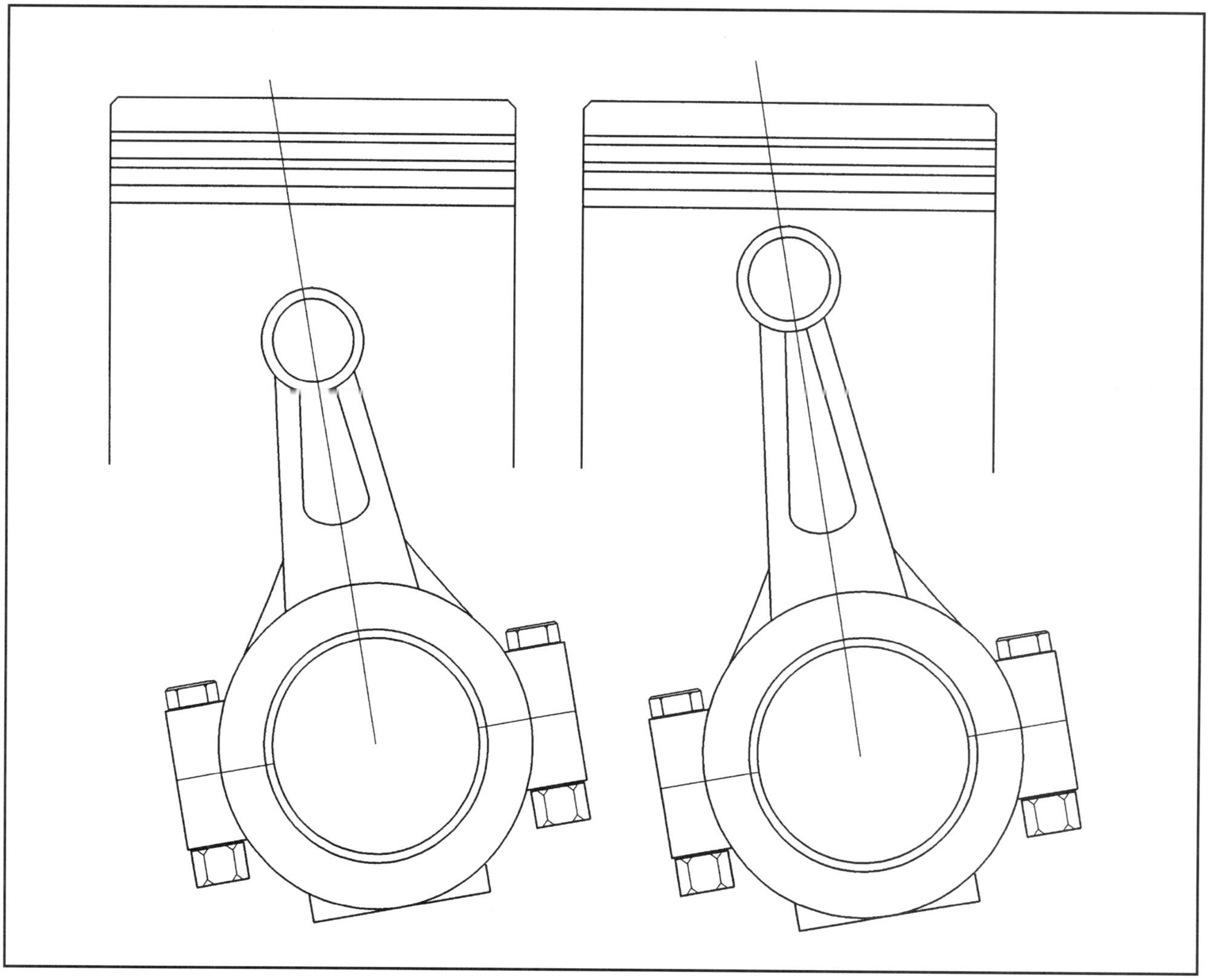

The piston on the left has the wrist pin located low on the skirt. This design places the point of maximum thrust close to the edge of the piston; this will cause the piston to rock violently. In the design on the right, the wrist pin is higher on the skirt, closer to the head of the piston. This reduces the lateral force, moving the thrust angle closer to the center of the piston. The tendency of the piston to rock is reduced. The tendency of the piston skirt and cylinder wall to wear is reduced.

best bet is to simply purchase a set of replacement performance pistons designed by someone who has already taken the best possible thrust angle into consideration when designing the pistons. The designer will also make sure that the design is not at the expense of a loss of compression ratio.

Oversized wrist pins

Since the wrist pin bore of the piston is much softer than the wrist itself, you will find that the wrist pin almost never wears significantly. The piston bore does wear and can be machined and fitted with a larger, or oversized, wrist pin. Let us look at the logic of this operation. A new set of pistons will cost in the neighborhood of $320. The machining of the pistons will cost in the neighborhood of $80 to $120. Now granted there is a big difference in these two neighborhoods; however, if the pistons have been in the engine long enough to damage or wear the wrist pin bores, then they have had a good, long life and deserve to be retired. Besides, would you rather assault the Al-Can Highway or Death Valley with fresh pistons or pistons that have had the wrist pin bore machined and an oversized pin fitted? It seems that when they will charge you $5 per mile for towing on the Al-Can, reboring may be false economy. On the other hand, if the engine you are working on is from a 1928 DeSoto 3.2 liter (ltr) side-valve six-cylinder, then almost any amount of machining could be justified.

Balancing the pistons

Rebuilding an engine is a really simple and fun procedure.

On more than one occasion I have managed to rebuild an engine in a day. The real challenge of rebuilding an engine is to take the time to create as perfect a machine as possible. The dynamic forces of the reciprocating components of the engine become amplified as the engine speed increases. An imbalance puts the crankshaft and other components under fluctuating stresses. These fluctuating stresses create a vibration; the vibration is energy being lost to rocking the engine instead of pushing the vehicle down the road. Bottom line: balancing the engine's reciprocating components will decrease the vibration and this will increase the power to the ground. Reach into your pocket and pull out a dollar bill. Now cut the dollar bill in half. Pick up one half. Note the weight. It is about one half gram. This is how closely the pistons should be balanced. In fact, this is how closely each of the reciprocating masses, piston, wrist pin, connecting rod, and bearing should be balanced.

Begin the balancing procedure by installing the wrist pin and its retaining clips, if applicable, into the pistons. Place each of the piston assemblies, in turn, on the scale. You will now remove metal from the heaviest ones to make them all weigh the same as the lightest ones.

Mount the first piston on a lathe. Use the lathe cutting tool to remove metal from the wrist pin bosses inside the piston. Repeat the process for each of the remaining pistons. Remember that you will still have the connecting rods to balance; therefore, a little extra exactness will mean less balancing work when the pistons are mated to the connecting rods.

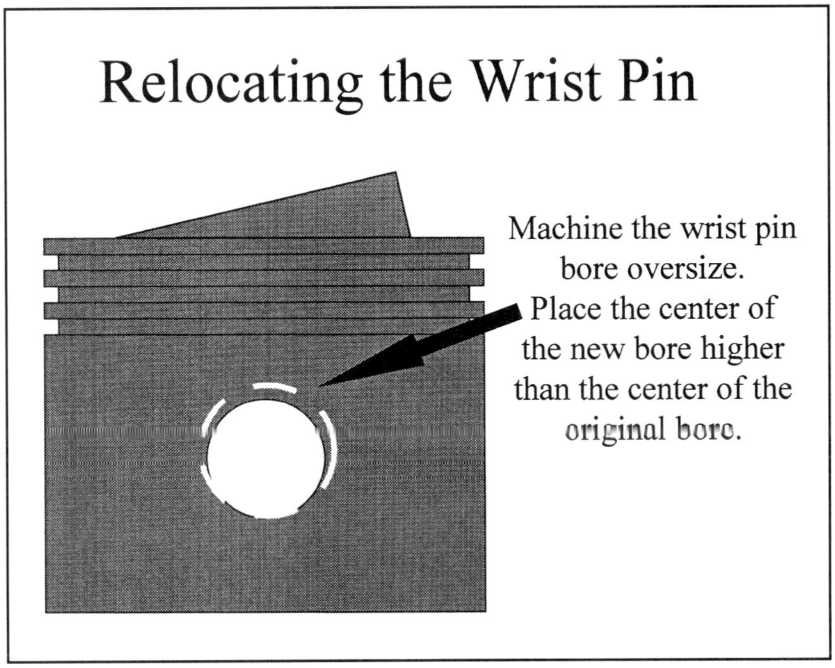

To improve the thrust angle, machine the wrist pin bore off center toward the head of the piston for an oversized wrist pin. This will locate the wrist pin closer to the head of the piston by 50 percent of the increase on the wrist pin diameter.

Even for an engine destined to haul mom and pop to the soda shop, balancing the pistons is a good idea. Balancing the reciprocating units, not just pistons but also connecting rods, can reduce vibration and increase satisfaction with the overhaul.

Chapter 3
Piston Rings

Square inch for square inch, the piston rings may well be the hardest working items in the engine. They slide up and down the cylinders at speeds approaching 46mph, and even more. They can be asked to accelerate from 0ft per minute to over 4,000ft per minute, then back to 0ft per minute 67 times per second and more. This is like accelerating from 0 to 3,000mph then back to zero again in one second. If that rate of acceleration could continue for sixty-two hours, the piston and rings would reach light speed. If you could build a car, or even a rocket, that could sustain this rate of acceleration, it would do you little good. Without inertial dampers, the occupants of the vehicle would be strained through the rear bulkhead of the vehicle like broth through a colander.

This rate of acceleration and deceleration subjects the piston rings to tremendous forces. But that is not all. Let's add to this the fact that several thousand times per minute the rings must also seal the forces of an explosion. In spite of all this, on

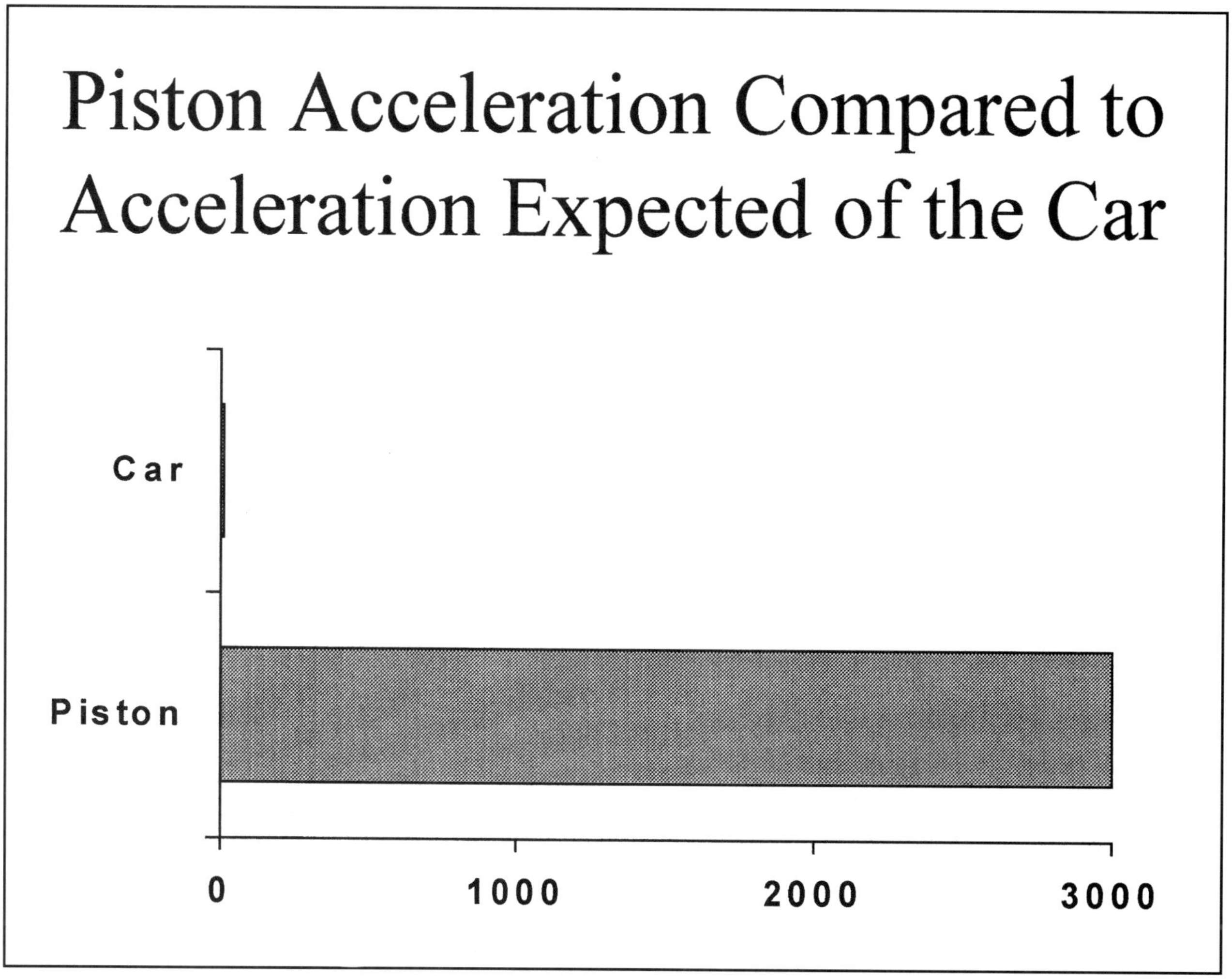

This chart is a comparison of what is expected of the piston and its rings and what we expect of the car in which the piston and rings are installed.

most Detroit iron or import "voiture d'elegance," these piston rings hang in there year after year reliably doing their job until finally, after 800 million cycles or more, they have the audacity to begin to bypass a significant amount of combustion gases or oil.

Like everything else in the world of engine rebuilding and performance modification, there are several different quality grades of piston rings. Let's face it, most rebuild applications are going to do little more than travel to the grocery store and maybe do a little bracket racing. For these applications, less expensive rings will do a good job and last a long time. If the goal is to get from point A to point B in the shortest possible time, then something a little tougher will be in order.

The typical spark-ignition engine will have three rings. The top two rings are gas sealing rings. The bottom ring is designed to reduce the amount of oil migrating up the cylinder walls toward the combustion chamber. The top rings are typically referred to as compression rings; the lower ring is called the oil ring or oil control ring.

Materials

Cast iron

Cast iron piston rings were the standard of the industry for decades. In the early 1970s, I was driving from Fort Worth, Texas, to Phoenix, Arizona. While passing near Van Horn, Texas, my little Fiat 850, which had been singing along at an impressive 75mph (almost legal in west Texas at the time), lost most of its reserve power. This little engine had been asked to work like an Indy 500 engine for the past 489 miles. Designed for the Italian market, a market where the typical trip is probably from the *domus* to the local *macellum*, the engine in this car was designed to go at best a few kilometers at a time. In west Texas, the distance from your ranchero to the nearest stockyard might be several hundred miles. I had asked an engine designed for short distance operation in a relatively cool climate to push my little car several hundred miles in the searing heat of the Permian Basin. The engine components were not up to the task; a compression ring on the number one cylinder broke.

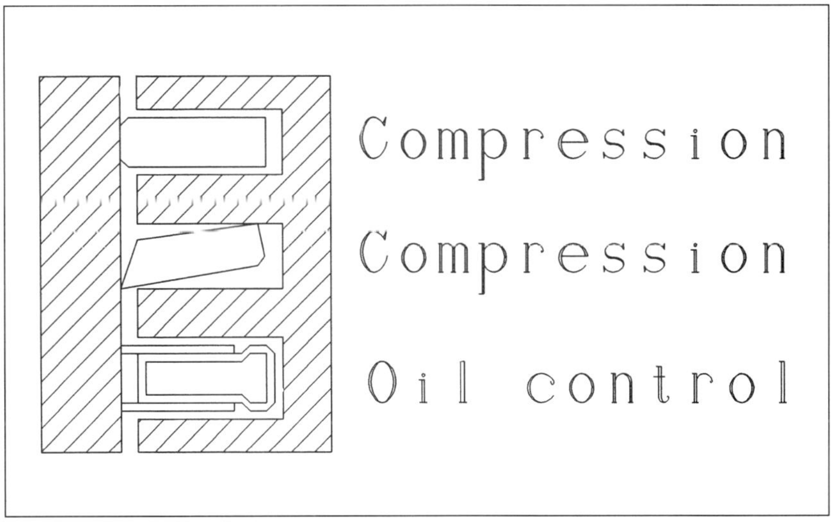

The typical spark ignition engine will have three rings. The top two are gas-sealing rings. The bottom ring is designed to reduce the amount of oil migrating up the cylinder walls toward the combustion chamber. The top rings are typically referred to as compression rings; the lower ring is called the oil ring or oil control ring.

The most common type of piston ring is simple cast iron. It features good durability and short break-in time. This cross-section illustrates the homogeneous nature of the alloy used in these rings.

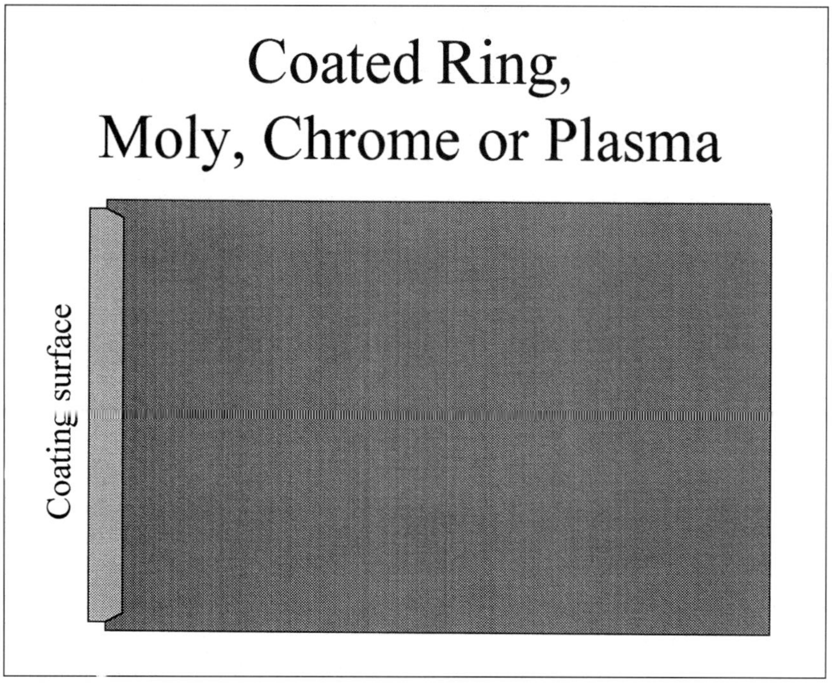

The primary advantage of moly rings is the natural porosity of the molybdenum dioxide coating. This porous layer absorbs oil and allows for extremely good lubrication of the cylinder walls without allowing oil to bypass to the combustion chamber.

Chrome rings are for those who have built an engine with perfect cylinder bores. Because these rings are so resistive to wear, they are intended to be used only in perfect cylinder bores. With lesser rings, a little time and a little patience with the smoking will let the rings seat.

The facing on plasma rings is a mixture of titanium oxide and aluminum oxide. This mixture is sprayed on the face of the ring. Now this may sound simple on the surface; however, these two substances are a lot like chalk dust. Imagine trying to spray paint with chalk dust and getting it to stick. After spraying the face of the ring, a plasma torch is used to heat the ring surface to between 30,000 degrees F and 40,000 degrees F.

Ring types

Cast iron piston rings are very durable but are brittle and can be damaged by combustion extremes such as detonation. Few modern applications use this technology once you look past the performance characteristics of a Cushman.

Moly

According to *The American Heritage Dictionary*, "molybdenum: Symbol Mo A, hard, silvery white metallic element used to toughen alloy steels and soften tungsten alloy, melting point 2,617 degrees C."

Many alloys used in high-performance automotive applications contain molybdenum. This element is used to harden and strengthen performance piston rings. During the manufacturing process, the standard cast iron ring is manufactured with a groove around its outer perimeter. As a final step in the manufacturing process, this groove is spray filled with molten molybdenum. As the molybdenum cools, it becomes molybdenum dioxide, a ceramic material much tougher and harder than the molybdenum itself.

The primary advantage of moly rings is the natural porosity of the molybdenum dioxide coating. This porous layer absorbs oil and allows for extremely good lubrication of the cylinder walls without allowing oil to bypass to the combustion chamber. Early experiments with this technology created a molybdenum dioxide layer that was extremely porous, so porous in fact that dirt and abrasives would be trapped in the coating and act as an abrasive. Today's moly rings have a porosity of less than 5 percent. Most modern engines feature a moly top compression ring from the factory.

Chrome

According to *The American Heritage Dictionary*, "chromium: Symbol Cr A, lustrous, hard, steel-gray metallic element, resistant to tarnish and corrosion and found primarily in chromite. It is used as a catalyst, to harden steel alloys and produce stainless steels, in corrosion-resistant decorative platings, and as a pigment in glass, melting point 1,890 degrees C."

Chrome rings are for those who have built an engine with perfect cylinder bores. Because these rings are so resistive to wear, they are intended to be used only in perfect cylinder bores. With lesser rings, a little time and a little patience with the smoking will let the rings seat. With chrome rings, what you have on the first day after an overhaul is what you have from that day forward. In most uses of chrome rings, the rings will easily last as long as or even outlast the rest of the engine.

As perfect as these rings may sound, no component is perfect. Moly will not bond, or weld, to the walls of the cylinder. Chrome will. Also, their use is not recommended for use with gaseous fueled engines. Do not laugh; I know a fellow in Dallas, Texas, who races a compressed natural gas dragster, quite successfully, and current EPA regulatory projections imply a requirement for alternative fuels by the turn of the century.

Plasma ceramic

Now these rings are really cool. The facing on the rings is a mixture of titanium oxide and aluminum oxide. This mixture is sprayed on the face of the ring. Now this may sound simple on the surface; however, these two substances are a lot like chalk dust. Imagine trying to spray paint with chalk dust and getting it to stick. After spraying the face of the ring, a plasma torch is used to heat the ring surface to between 30,000 degrees F and 40,000 degrees F. This is actually warmer than an August afternoon in Phoenix. At these temperatures, the ceramic titanium oxide and aluminum oxide powder will adhere to the surface of the ring. This process gives the cylinder wall mating surface of the ring a ceramic layer of 0.004in to 0.008in thick.

Pressure-balanced rings

The pressure-balanced ring was designed to allow combustion pressure to sneak behind the ring. This pressure behind the ring forces the face of the ring against the cylinder wall. Typical of this type of ring is a narrow mating surface with the cylinder wall. The narrow mating surface allows for reduced friction as well as better performance and fuel economy. Narrow mating surface, extra machining for gas passage, it all adds up to extra expense. For maximum performance pressure, balanced moly or chrome rings are ideal.

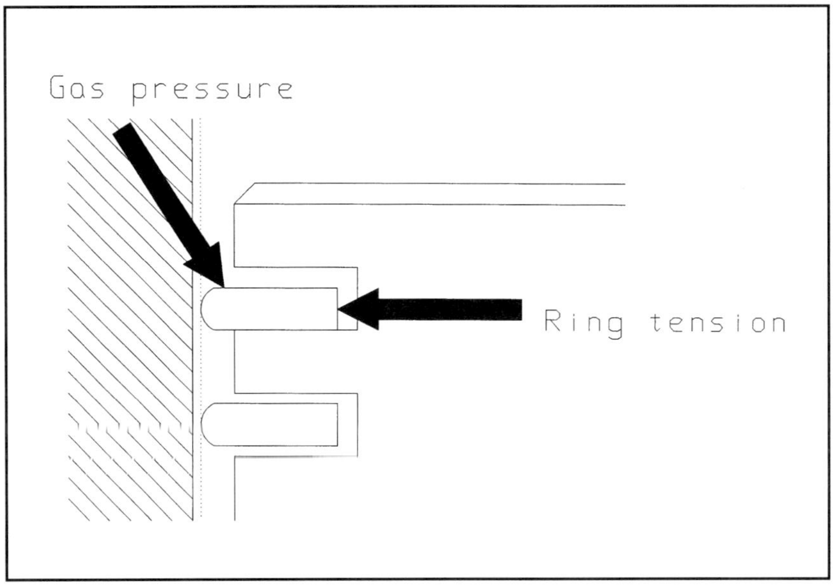

Pressure-balanced rings are designed to use the force of the combustion gases to ensure a high-pressure contact between the ring and the cylinder wall.

Selecting the type of rings best for you

As so many things are, what is ideal for one application is not necessarily ideal for another. The three primary considerations are the general quality of the work being done during the engine rebuild, purpose for rebuilding the engine, and cost. Chrome rings should not even be considered if the engine cylinders are not going to be trued to a perfect circle. The hard, tough surface of chrome rings will prevent them from seating in properly for thousands of miles, if ever. Nevertheless, chrome rings are the rings of choice in an engine "blueprinted" for power where price is an issue.

Moly rings will seat in nicely in a few hundred miles. However, a common mistake made by the amateur and, sad to say, professional rebuilder is failing to make sure that the cylinders are sufficiently round. As the engine runs during its first life, the cylinder walls are slowly machined into an oval by the rings. Piston offset is the primary agent responsible for this. If these walls are not retrued during an overhaul, the second life of the engine will be short and unsuccessful. While the precision of this truing need not be as exact as they would if the rings destined for this second life were chrome, no engine rebuild should include putting rings into old, imperfect cylinders.

In an engine where power and durability are important, like a racing engine, where the dynamic and explosive forces of compression and combustion are very much active, the plasma ceramic ring may be the most appropriate. The expense of these rings makes them impractical and unnecessary in an engine destined to deliver the user to the local grocery. These rings typically cost $150 to $250 per set. This compares unfavorably to $30 for a set of cast iron rings and $50 to $75 for a set of moly or chrome rings. TRW claims these rings are five times stronger than a stock cast iron ring. Engines that will be used on the drag strip, especially those that are not naturally aspirated, those that are turbocharged or supercharged, would benefit from the extra expense involved in plasma ceramic rings.

Ductile iron vs. plain iron

Types

As has already been suggested, there are several ways to

classify types of rings. The broadest classifications are by the tasks the ring is assigned. The basic task of all rings is to seal something. The categories of "somethings" are compression and oil.

Compression rings

Torsional twist rings

Now for years I pictured the compression ring as a symmetrical sealing ring. I imagined that it sat more or less statically in the ring groove of the piston, much like the sealing rings of a transmission shaft. I saw them sitting in place and merely sliding up and down the cylinder passively preventing blow-by and controlling oil. The reality is much different from my limited imagination.

A torsional twist ring features a chamfer on its top inner surface. It sits in the ring groove with a twist that forces a line of contact between the upper edge of the ring and the back of the ring groove. When there is a minimum load on the rings, such as during idle and under light coast conditions, this attitude of the rings helps to prevent gases from passing around the back side of the ring during combustion and prevent oil from passing around the back side of the ring during the intake stroke. This clarifies for me why the ring installation instructions always indicated that the chamfered inner edge of the ring should go toward the top. I read this in the installation instructions; I obeyed the installation instructions; but like so many of my generation, I always wanted to install the rings upside down just because it defied the ill-explained instructions. Question authority.

During acceleration, as the combustion pressures reach their greatest levels, the torsional twist ring is forced against the bottom of the ring groove. This seals the combustion and prevents excessive blow-by. During deceleration, the torsional twist ring is forced against the top of the ring groove and assists the oil rings. Installing these rings upside down, listening to the temptations of my youth, would have led to low compression and high oil consumption.

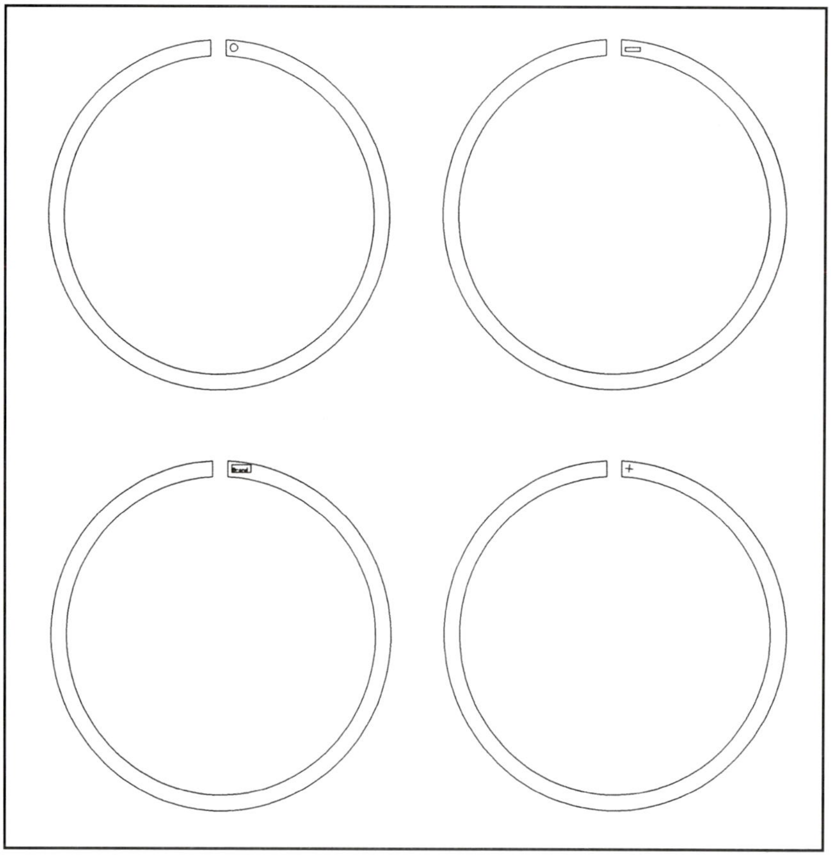

The design of many piston rings is such that there is a specific side that must be up when installed. The "up side" is usually marked with a small circle or other symbol. Be sure to read the instructions that came with your new set of rings. On rare occasion you may find a manufacturer that wants this symbol installed facing down.

Reverse twist rings

Reverse twist rings might have served to mollify the rebellious tendencies of my youth. These rings feature a chamfer on the lower inner edge of the ring. In reality, this would have satisfied no inner craving for rebellion. With these rings, the instructions clearly indicate that the chamfer should be installed down; therefore, I would have wanted to install them with the chamfer up. A frustrating dilemma.

In addition to the chamfer being on the lower inner edge of the piston ring, it is tapered at an extra sharp angle that forces a contact line between the lower edge of the ring and the cylinder wall. The reverse twist maximizes oil control but is less effective in controlling blow-by gases. A common setup would feature a torsional twist ring as the top ring and a reverse twist ring as the second ring.

Barrel-faced rings

The barrel-faced ring does not twist but rather remains relatively square in the ring groove. The slight 0.001in curve on the face of the ring allows for any misalignment in the ring grooves. This curve also allows for a high-pressure, though very narrow, contact surface between the ring face and the cylinder wall. Barrel-faced rings often feature a moly or chrome face and may be chamfered to gain some of the benefits of either the torsional twist or reverse twist ring.

Low-tension rings

Low-tension rings, as their name implies, minimize the tension at the contact point of the ring with the cylinder wall. This design allows for slower cylinder wear, a feature that may be obvious. A little less obvious is the fact that the low tension reduces energy loss. Their main disadvantage, like that of the chrome ring, is that the cylinder must be in near perfect machined condition. Maximum taper is less than 0.006in and maximum cylinder out of round is 0.005in. These are considerably closer tolerances than can be hoped for from your typical dingleberry hone job.

Ring combinations

In reality, a package of rings will generally include two different compression ring types. A top ring of the moly barrel variety and a reverse twist second ring would make up a typical set.

Oil control rings

The most common oil control rings are three-piece assemblies. A thin top and bottom spring steel or chrome ring is separated by a springy stainless steel separator or expander. There are some low-performance applications that use a one-piece oil control ring, but this is virtually unheard of in modern performance engines.

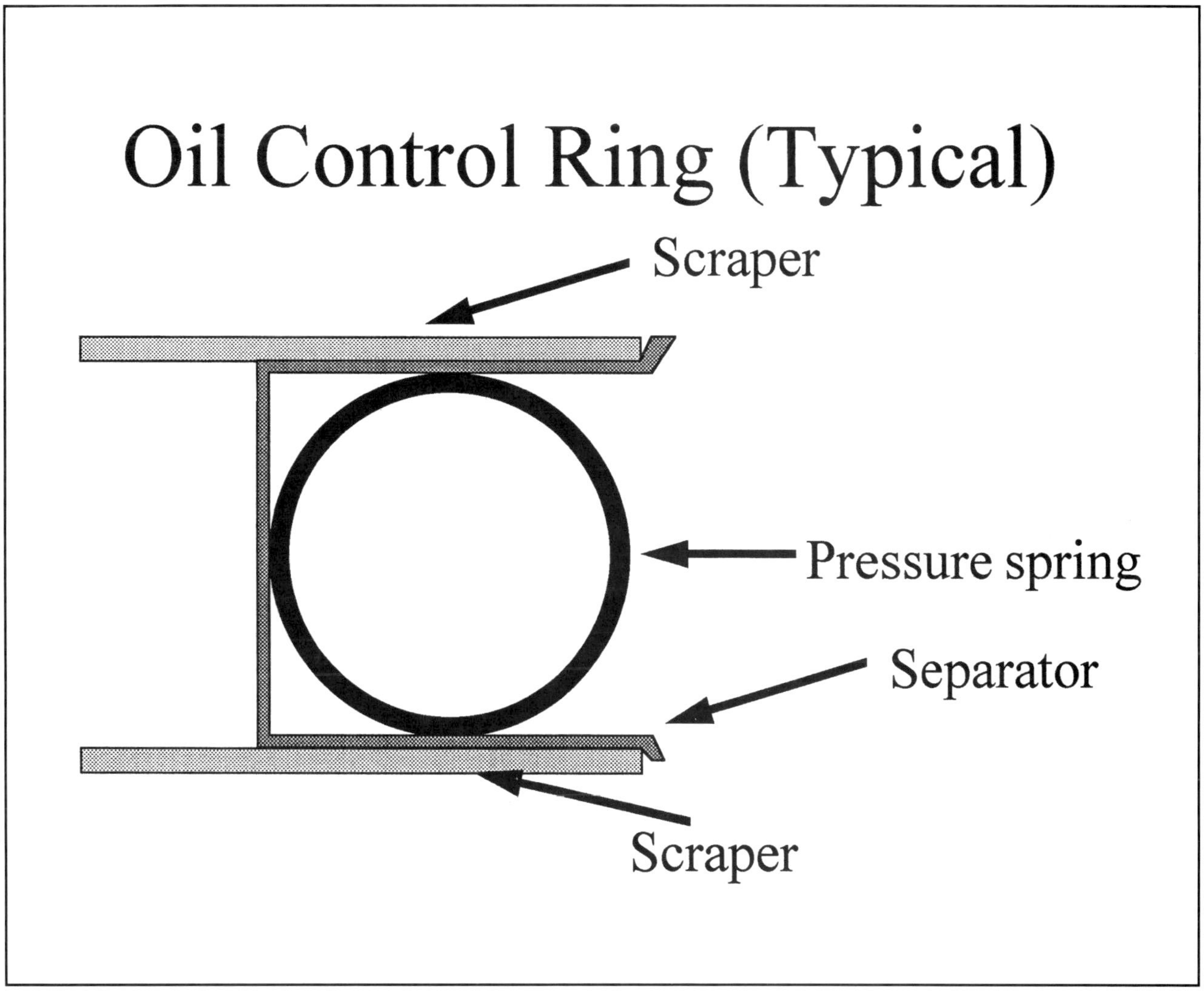

Oil control rings feature two sharp, thin rings or ring edges. These sharp edges are designed to scrape oil off the walls of the cylinder. These sharp-edged rings are separated by a spring steel separator. In the example shown above, the separator is forced away from the inner ring groove wall by a spring. This ring can come in variations containing all of the three above functions: one-piece, three-piece, and four-piece. This is a four-piece ring.

Measuring the ring gap clearance is one of the operations where some of the greatest powers of performance engine building sorcery is required. The idea is to calculate the thermal expansion characteristics of the rings around their circumference and trim the rings so the gap will be almost nonexistent when the engine is at operating temperature. Here the gap of an old and successful rebuild is measured prior to an overhaul.

The function of the oil control ring is to form a tight seal with the walls of the cylinder. This tight seal makes the oil control ring the greatest offender of all the rings when it comes to power loss as a result of friction between the rings and the cylinder walls. Engine vacuum rises during deceleration. Actually, during the intake stroke of an engine that is decelerating, the receding piston is creating a low pressure in the cylinder. During acceleration or cruise, air is supplied to the cylinder through the open throttle valve, or butterfly. During deceleration the amount of air that can pass through the throttle bore is extremely limited by the closed throttle valve. This causes the low pressure of the combustion chamber to draw air from any available place. The part of the engine that has both direct connection to the combustion chamber and high pressure is the crankcase. This is okay in general; air drawn from the crankcase would not contain fuel and therefore would not contribute power to the engine. The engine would still decelerate; however, the air from the crankcase is laden with a mist of engine oil. This mist would enter the combustion chamber and be burned, creating a smoke screen behind the car that would make 007 proud.

Ring gap clearance

This is a science unto itself. My best suggestion is that you assign decisions concerning ring gap clearance to an individual who you can blame later if things go wrong. As the engine runs, the rings are heated both by friction and by combustion. The heated rings expand. If there were no gap between the ends of the rings, they would soon expand to be larger than the cylinder bore. This creates a potentially disastrous condition called negative clearance. To facilitate thermal expansion of the rings, a gap is left between the ends of the rings. When measuring the end gap of a piston ring, always use the engine manufacturer's specifications. If high-dollar, "pricey" rings are chosen for use, follow the recommendations of the piston ring manufacturer.

If such specifications are missing or unavailable, a typical spec for this gap is 0.0035in per inch of cylinder diameter. Using this spec, an engine with a 4in bore should have a gap of 0.014, fourteen thousandths of an inch.

Ring end gap is only highly critical in engines on applications where the very last ounce of power is critical. For most engine overhauls, it is far better to have excessive end gap than it is to risk not having enough. In a performance engine, every thousandth of an inch of excessive ring gap translates instantly into a power loss; therefore, specs from the ring manufacturer become critical. Only they know for sure the true expansion rate of the ring.

Many high-performance rings are sold with a negative end gap. This allows the installer to set the ring gap as he or she feels is best for the engine he or she is building. A special cutter is required to cut these rings. If you do not already possess such a cutter, purchase it when you buy the rings. Place each of the piston rings in the cylinder in which it is to be used. With a feeler gauge, measure the gap between the ends of the ring. Performance rings often have no end gap or overlap when installed. K-D Tools makes a special cutting/filing tool that can be used to cut/file the rings to the perfect gap and yet ensure that the end surfaces of the rings are perfectly parallel. After the rings are cut, the rough edges should be removed by hand filing. Consult the ring manufacturer for the optimum ring gap for your application and usage.

Ring side clearance

Ring side clearance should also be checked. A ring that fits too snugly in its groove may become jammed in the groove and fail to seal properly when the engine gets warm. Place a feeler gauge in each of the ring grooves with the ring that is destined for that groove installed. Progressively increase the thickness of the feeler gauge until a snug fit is obtained. Ring side clearance should be 0.002 to 0.004in.

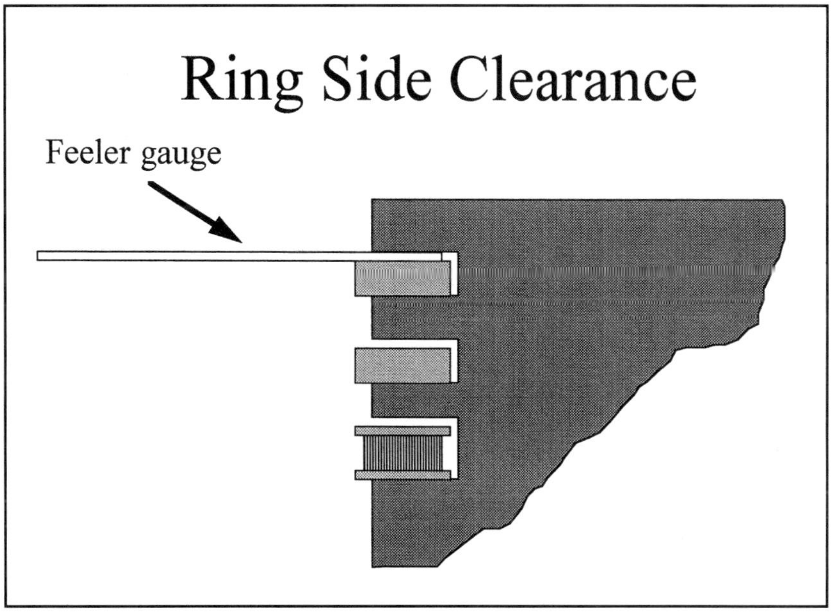

Ring side clearance should also be checked. A ring that fits too snugly in its groove may become jammed in the groove and fail to seal properly when the engine gets warm. Place a feeler gauge in each of the ring grooves with the ring that is destined for that groove installed. Progressively increase the thickness of the feeler gauge until a snug fit is obtained. Ring side clearance should be 0.002 to 0.004in.

Recommended ring end by engine usage (represented engine has a 4in bore):

Ductile iron top rings	**Inches of gap per inch of diameter**
Supercharged/injected nitro engine	0.022-0.024
Supercharged/injected alcohol engines	0.018-0.020
Supercharged/injected gasoline engine	0.022-0.024
Oval track (rectangular ring) carbureted engine	0.018-0.020
Oval track (head land rings) carbureted engine	0.024-0.026
Oval track (pressure back rings) carbureted engine	0.020-0.022
Modified drag racing carbureted engine	0.018-0.020
Stock drag racing carbureted engine	0.016-0.018
Street carbureted engine	0.016-0.018
Plain iron second rings	
Supercharged/injected nitro engine	0.014-0.016
Supercharged/injected alcohol engines	0.012-0.014
Supercharged/injected gasoline engine	0.012-0.014
Oval track carbureted engine	0.012-0.014
Modified drag racing carbureted engine	0.012-0.014
Stock, Super Stock drag racing carbureted engine	0.010-0.012
Street carbureted engine	0.010-0.012

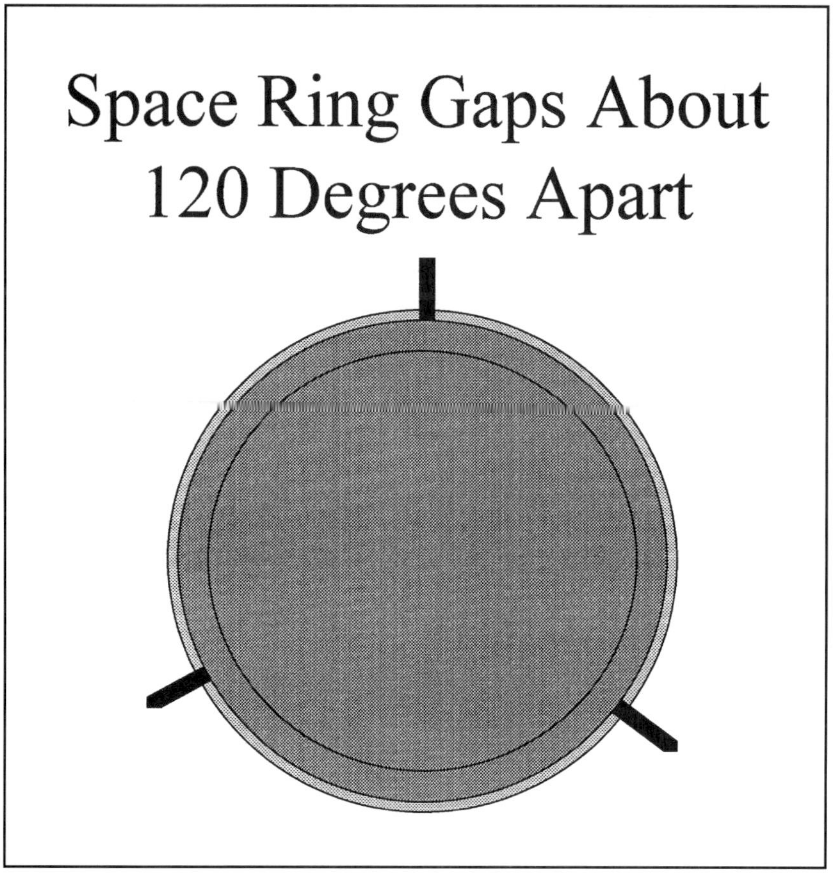

After the ring grooves have been cleaned as necessary, the rings should be arranged at 120 degree intervals around the piston prior to installation in the block. This ensures best compression and lowest chance of oil burning during the break-in period. I should tell you that the rings will rotate as the engine is run and will not be arranged at 120 degree intervals when the engine is taken apart again.

Left
The block is properly prepared for the installation of the new rings when there is a distinct 45 degree crosshatch apparent in the cylinder. This can be achieved either by honing with a flat stone hone, with a berry-ball hone, or by having it done by your local machine shop.

Installing rings on the pistons

Follow the supplied instructions closely. If you have not done so already, be sure that the ring grooves are clean and clear of debris and carbon. Use a special tool called a ring groove cleaner. However, for those who have good health insurance and a first aid kit, a broken ring can be used to scrape the carbon from the grooves. Extreme caution should be used when using a ring to clean the grooves. The broken ring is sharp and can leave nasty cuts. Wear gloves.

With the ring grooves clean, install the new rings on the pistons. Stagger the rings before installing the pistons into the cylinder. Staggering will provide a better seal during the early stages of running the engine. As the engine runs, the rings will often rotate. Some applications employ a pin in the ring groove to the piston to prevent rotation. It is essential that the rings on these pistons be staggered at 120 degree intervals. The rings are usually identified top, middle, and oil control by the papers covering them in the box. There will also be a mark on the two compression rings indicating which way is up. This mark may be a dot, a bar, or even a corporate logo. In a few instances the rings may be unmarked; there will be instructions in the box.

Two techniques are available for the actual installation of the rings. Start one end of the ring in the correct groove. Rotate the ring around the center of the piston and it will practically fall into place. Plan on breaking an occasional ring using this technique. Piston ring installation pliers can be purchased for only a few dollars at most auto parts stores. Using these pliers virtually eliminates the possibility of broken rings.

Honing the cylinders in preparation for new rings

Honing the cylinders is an operation that can be done by anyone. An engine cylinder hone can be purchased for as little as $20. Honing is done to remove the glaze from the cylinder walls. There is a common, but mistaken, belief that honing is used to true the cylinders. Perhaps that belief comes from the common practice of honing the last few thousandths of an inch when the engine cylinders are bored. The fact is that as the engine is run over the years, the old piston rings polish the walls of the cylinders. Honing breaks up this glaze and provides the rings with a slightly abrasive surface to help seat them.

A correct honing technique will produce a cross-hatch pattern on the cylinder walls. Place the hone in a low-speed electric or air-powered drill motor. It should be noted that in order to produce a 45 degree cross-hatch with an air-powered drill motor rotating at 20,000rpm, your arm must be capable of reciprocating about 330 times per second. So if you are not a special government project with a $6 million price tag, make sure you use a slow drill speed. Hold the drill and hone vertically and move it in and out in slow even movements. Continue the process until there are conspicuous 45 degree cross-hatch marks in each of the cylinders.

Having the engine block professionally honed at a machine shop is advisable. A well-equipped machine shop will have an automated honing machine that will ensure proper cylinder wall alignment and cross-hatching.

After the rings have been installed on the pistons, there is still the challenge of slipping the pistons into the cylinders in spite of the fact that the rings do not fit. Each ring is shaped so that it is larger than the cylinder in order to maintain firm contact between the ring and the cylinder wall. A gap in the ring allows it to be squeezed together so it will fit in the cylinder. The open end of the rings must be squeezed together in order to get the pistons into the cylinder. This is the job of the piston ring compressor. While any piston ring compressor will work for most engines, some engines require a special compressor. With most domestic and foreign engines, the pistons go in from the top. A few engines, such as the old Volkswagen air-cooled engines, require the pistons to be installed from the bottom. On these applications, the connecting rod is attached to the crankshaft before it is installed into the cylinder. A standard ring compressor could not be removed once the piston is installed. Although there are ways around this problem, there is a special ring compressor available, one designed to come apart for easy removal.

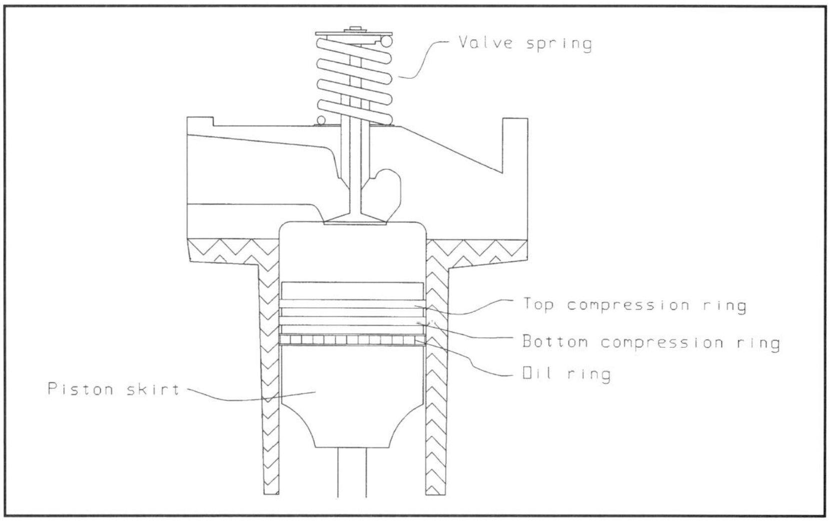

The piston and rings should form a sealed pneumatic unit when properly installed. If everything is installed correctly and after the engine has been run a few hours, the percentage of cylinder leakage should be less than 5 percent.

Chapter 4
Connecting Rods

The job of the connecting rods is to connect the power generated by the explosion of the air and fuel in the combustion chamber above the pistons to the crankshaft. Like the pistons, the connecting rods can either be cast or forged. Forged pistons are stronger and more precisely manufactured; therefore, they are normally reserved for performance-oriented engines. Connecting rods for consumer-oriented cars are made of cast iron. Cast iron connecting rods represent an enormous reciprocating mass. This mass robs the engine of precious power and top rpm. Racing engines and other high-performance engines use aluminum connecting rods. The 1950s and '60s saw the introduction of aluminum connecting rods on performance engines. This allowed higher maximum rpm and less parasitic loss of power to the reciprocating movement of the connecting rod mass.

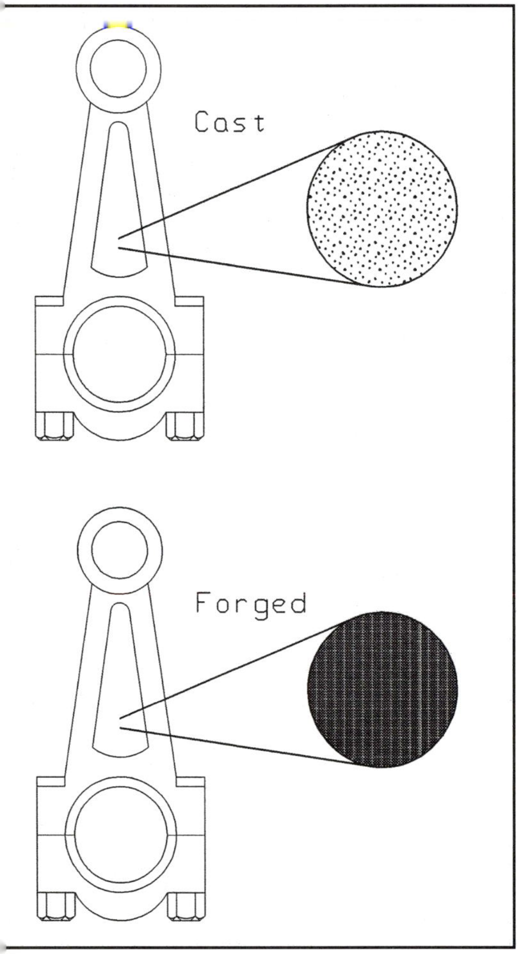

There are two primary materials that connecting rods are made from, steel and aluminum. Steel rods are used in applications where durability is required. This would include the family car. Aluminum rods are used in applications where low reciprocating mass is desired, such as racing engines. Both steel and aluminum connecting rods are available in cast and forged varieties. The forged connecting rod has a denser structure and, therefore, greater strength.

Each connecting rod has a big end and a small end. The small end fits into the piston where the power is transferred from the piston to the connecting rod by means of a wrist pin. The wrist pin may be pressed into the piston and floating on the rod. It may be pressed into the rod and floating on the piston, or it may be full floating in both the piston and the connecting rod.

The big end of the connecting rod joins the crankshaft. Around the inside of the big end is a replaceable bearing shell. Due to the enormous stress exerted on these bearing shells, they are among the engine components most prone to failure.

Materials

There are two basic materials that have been used in modern connecting rods, steel and aluminum. Steel connecting rods are still the industry standard for factory OEM (Original Equipment Manufacturers) engines. These engines are designed to take mom and pop to the bingo parlor or to provide the illusion of speed and performance, while making sure that even with the roughest abuse the engine will at least outlast the 50,000- to 70,000-mile warranty. Connecting rods connect the power of the controlled explosion on top of the piston in the combustion chamber to the ground. Since each of the cylinders fires consecutively, each of the connecting rods is individually responsible for moving several tons of iron, steel, aluminum, glass, and plastic down the road. A typical steel connecting rod features an "I-Beam" construction. This maximizes the strength of the rod while minimizing the weight. These steel connecting rods offer extreme durability in the roughest and most long-term service.

Performance connecting rods (remember that performance is a term that can mean either speed or economy) are generally made of aluminum alloy. My first experience with aluminum connecting rods was very positive. I was in trade school; as part of the course, we were required to do a real overhaul on a "real customer's" car. I was blessed, that would be a pleasant way to put it, with an early '60s Plymouth Valiant. The engine made no noise beyond the normal engine running noise during our initial inspection. Upon tearing the engine down, we found that all of the bearings in the big ends of the connecting rods had spun. In spite of this, the "customer" had driven the car in to have the engine overhauled only because the school would do it cheaply and she wanted to keep the car, not because she had symptoms that would warrant an overhaul. Although most of the engine, including the cylinder head, had to be replaced, only two connecting rods required replacement. The tough part of the job was convincing the parts suppliers, including the dealers, that such an engine ever existed. Even the doubters among you must admit that this illustrates the durability of aluminum connecting rods.

Aluminum connecting rods offer less inertial mass as their primary advantage. Lighter than steel, they can significantly improve acceleration. It is much easier to accelerate the reciprocating speed of a lighter component than a heavier component. Lighter connecting rods also allow lightening of crankshaft counterweights; this reduces the rotating mass.

Casting vs. Forging

Okay, here we have another of those terms that are thrown around as though everyone knows what they mean. "Yeah, I rebuilt my engine, put in a 3/4 cam, forged pistons, and forged rods." Great, what is a 3/4 cam? Is it a V-8 camshaft with four lobes rounded off? This term is loose slang for a mild grind on a performance camshaft. The term "forged," however, is much more specific.

Casting of iron goes back to the bronze age, as long ago as 6,500 years. The methods used to cast metals have changed little since. Casting of iron is a technology that dates from about 3,000 BC. Steel is a relatively new development. Basically, steel is an alloy of iron and carbon. Its production dates from only the fourteenth century. Casting of steel dates from only the 1800s.

When bronze, iron, steel, or any other metal is cast, it is done with a mold. The molten metal is poured in and allowed to cool. This creates a fairly strong, although often poorly structured, metal. Think about the structure of sandstone. There are great mountains of sandstone around the world; they hold up relatively well against erosion and the other forces of nature. Sandstone, however, is brittle and breaks easily when subjected to

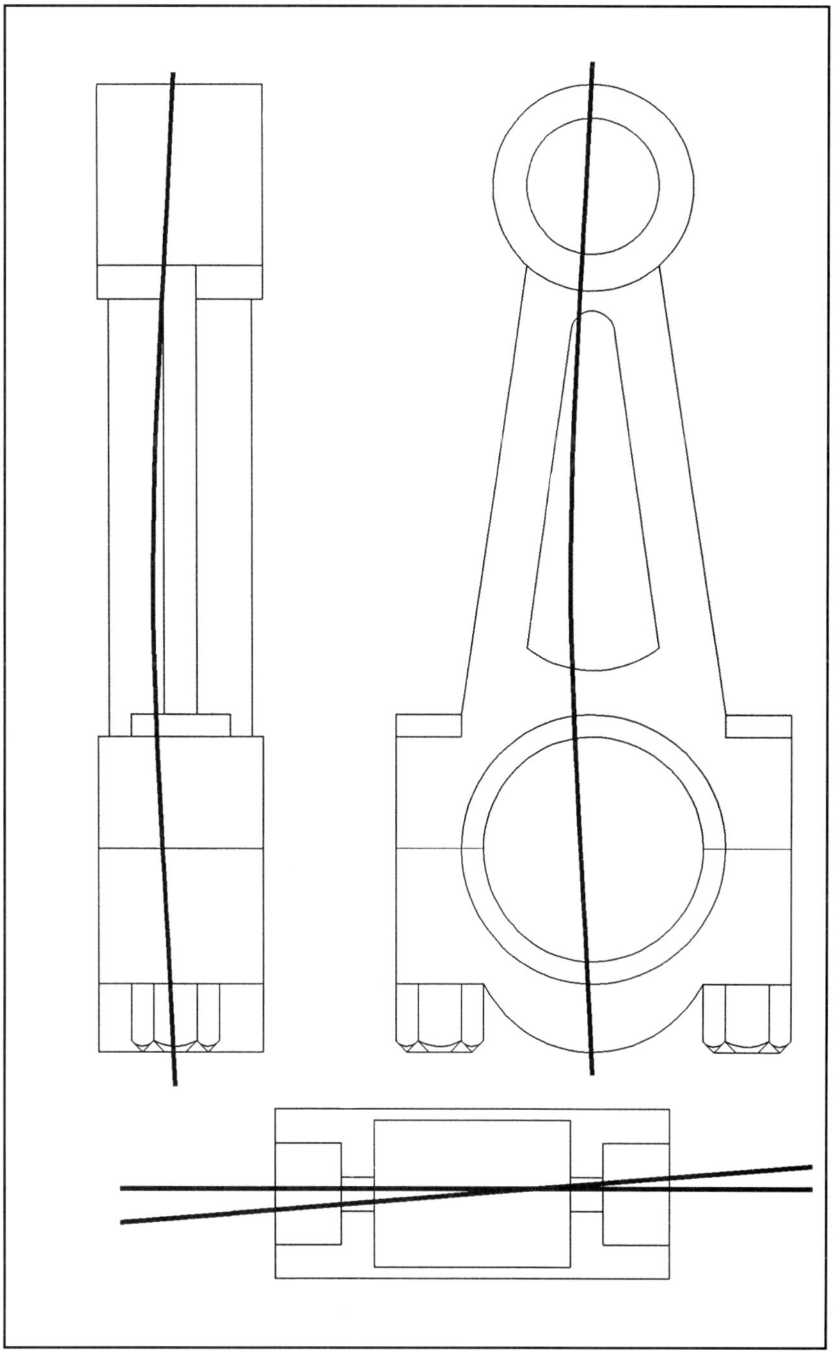

Connecting rods that are to be reused should be checked for cracks, heat damage, nicks and burrs, residual magnetism, as well as bending and alignment. This drawing shows the critical axes for bending and alignment.

lateral, shear, forces. Forging hammers the metal into a uniform, well-structured material. This is similar to the process that transforms sandstone into quartzite. Quartzite is a much more homogeneous stone, almost like a gem. It is far more resistant to weathering and the stresses of nature and has a much greater shear strength. Forged is to cast as quartzite is to sandstone. A forged connecting rod will therefore have a much higher tensile, or stretching, strength and a much greater lateral, or shear, strength. Aircraft connecting rods are almost always forged.

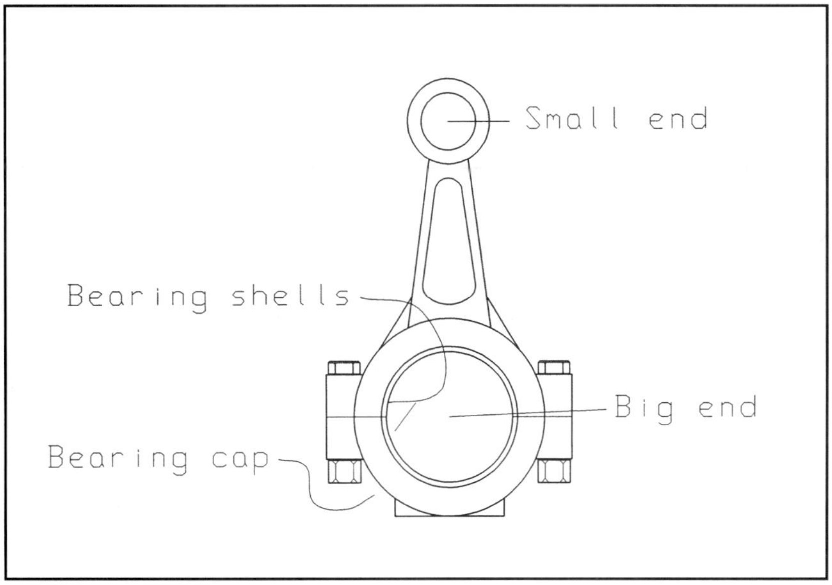

The standard connecting rod found in passenger cars and light-duty trucks is the cast steel connecting rod.

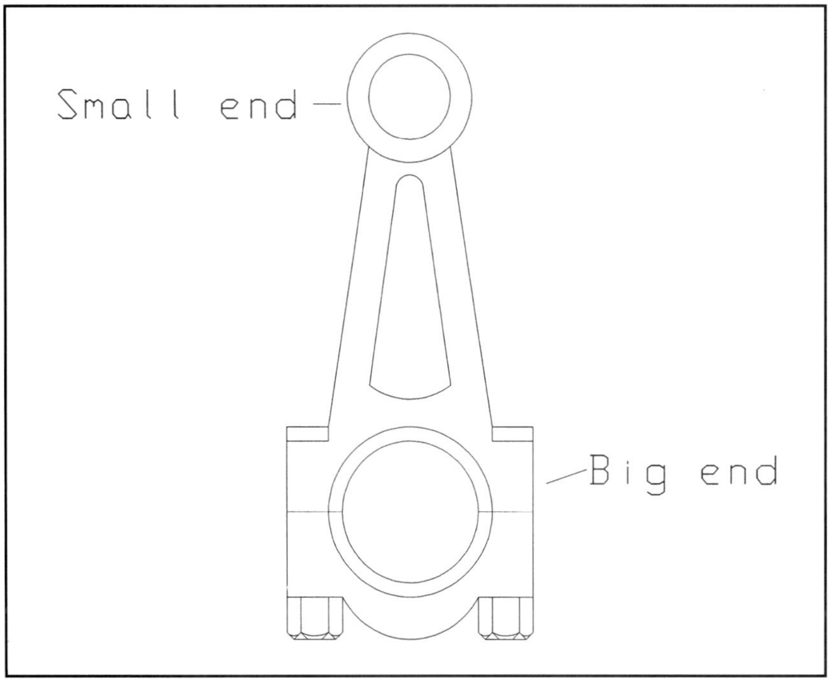

Performance engines, especially those where it is assumed teardowns and inspections will occur frequently, often use aluminum connecting rods.

Checking connecting rods

If you plan on reusing a set of connecting rods, especially if the rod has been in a racing engine or an engine that has suffered a catastrophic failure, there are several things for which it must be tested. These are: cracks, bends, heat damage, nicks and burrs, residual magnetism, and alignment.

Cracks

I guess it would be obvious to most anyone that installing a cracked connecting rod will doom an engine to early and perhaps catastrophic failure. The question is, how can you ensure that the rods you are installing do not have a hidden or microscopic crack? You can't. Even a new connecting rod could have an undetectable manufacturing flaw that could cause the engine to disassemble itself in a matter of milliseconds. The best we can do is reduce this likelihood through careful inspection.

Visual

This is where the "Star Trek" character Data or maybe the $6 million man would come in handy. Cracks that may damage the connecting rod and, therefore, the engine can start life extremely small. A good visual inspection requires a well-lit work area, a good magnifying glass, and a lot of patience. In fact, it requires a good eye. In every shop and every family there is a person who can find things no one else can find, a person who can see things no one else can see; this is the person who should do the inspection.

Magnafluxing

Even the most discerning individual can miss cracks that

A quick visual inspection of this connecting rod revealed no problems or defects.

give themselves away only as the slightest discoloration. More scientific methods are available, methods that can even detect cracks hidden below the surface. This technique is called magnafluxing. The connecting rod is placed on a powerful electromagnet. Fine iron powder is sprinkled on the connecting rod. The magnetic field will align the iron powder in a very distinctive pattern. A crack will interrupt this pattern. When the technician sees this interruption in the evenness of the powder, he knows that a crack is present.

What about aluminum?

I have just checked all the parts and tool catalogs in my little office here in Fairbanks,

Closer inspection revealed that the bearing shell had been wobbling in the big end of the rod. This was likely because the rod had been stretched when the engine was over-revved at some point during its life.

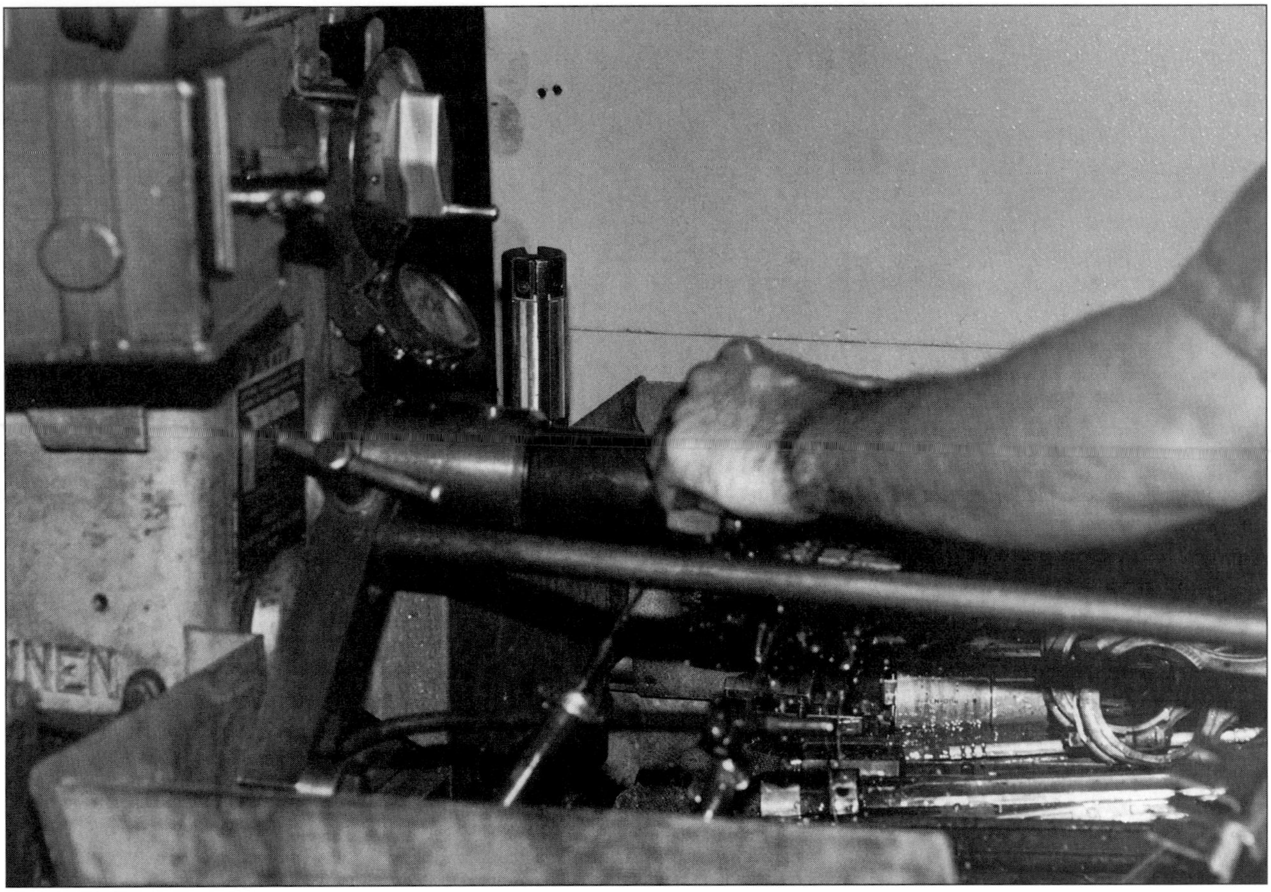

Damaged or bent connecting rods can often be repaired by an automotive machine shop. Here a machinist is using a lathe designed specifically for the repair of connecting rods.

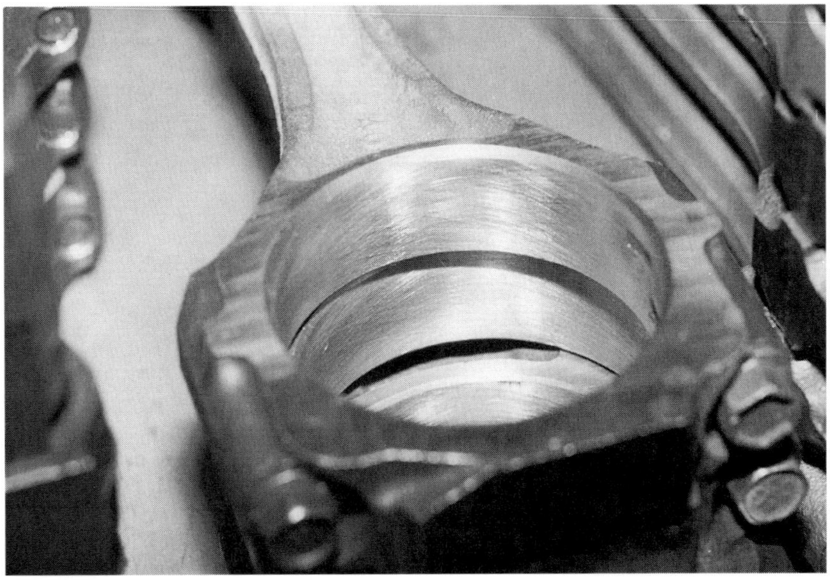

When a stretched or damaged connecting rod is repaired, the ends of the bearing caps are trimmed, the rod cap is bolted on, and the big end is remachined. These connecting rods have been repaired and are ready for installation. Notice the casting ridges on the stack of connecting rods to the right. These ridges offer a starting point for stress fractures and cracks.

Alaska. I cannot find any reference to an aluminum magnet; therefore, we must operate under the assumption that such a device does not exist. There are ways of checking aluminum components for cracks, although these methods cannot locate cracks hidden below the surface like magnafluxing can.

Dye penetrant

Imagine that you are traveling the Al-Can Highway in early April. The roads are clear, but the tourists have not yet invaded in waves. You pull into a small cafe in the Yukon Territories and ask for a cup of coffee. The coffee is strong, but you need it. As you get close to the bottom of the cup you see a narrow but long crack in the porcelain. This crack is brown. Your coffee, in fact the coffee of the last thousand users of the cup, along with their viruses and bacteria, have stained the ceramic material exposed in the crack. This is exactly how dye penetrant works. The crack exposes the relatively coarse and porous inside of the connecting rod to the dye. The dye soaks into the porous inside, but is easily cleaned off of the relatively nonporous outside.

The procedure involves cleaning the surface of the connecting rod with a colored dye. After allowing the dye to dry, as per instructions on the can, clean the dye off of the rod. This usually involves the use of a solvent. This dye on the outer surface of the rod will clean off easily; the dye in the crack will remain, making the crack visible.

So what if it is cracked?

In a phrase? Replace it. Okay, you are building up an engine from a 1948 Studebaker and you want to keep everything original. After extensive searching you are unable to find anyone with a good connecting rod for this engine. There are master machinist/welders who can repair the rod, but these folks are extremely rare so be careful; not everyone with a pair of goggles and a bottle of acetylene is a master. Ask several of the local racers and aircraft mechanics in your area for recommendations and go with the most recommended.

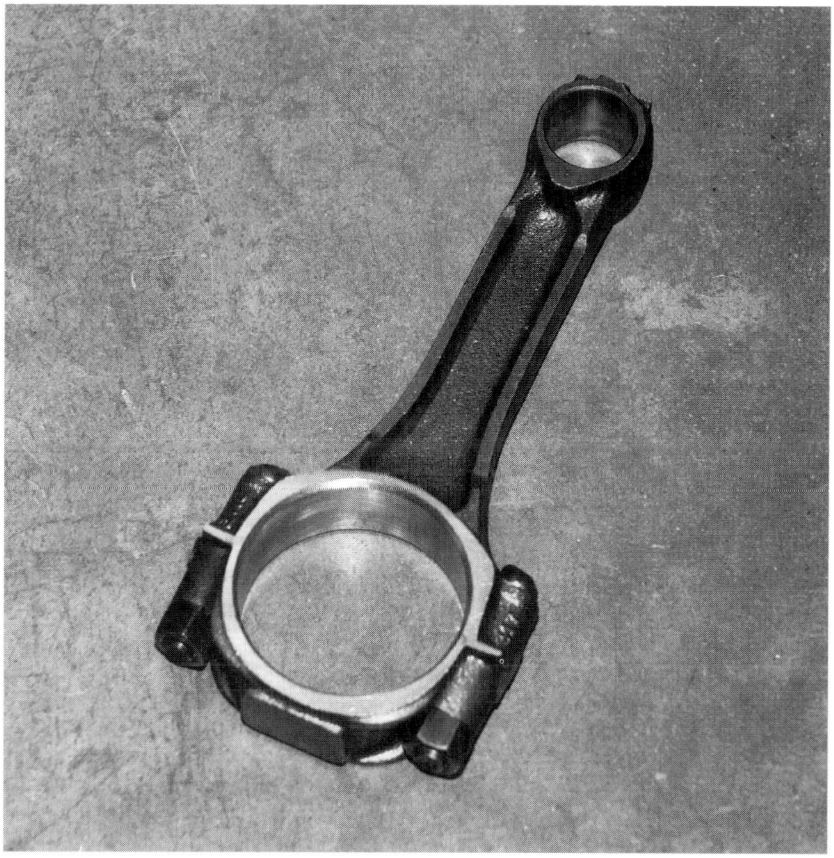

The flat place on the big end of the connecting rod offers an ideal spot to remove weight without weakening the rod. This is the place to file off metal to balance the connecting rods.

Bends

Remember the Volvo with the bent connecting rod I referred to earlier? That engine had a connecting rod that was shaped like a pretzel. This is an extreme and obvious example. This engine also had other evidence of a serious engine failure. The dangerously bent connecting rod is the one with the diminutive bend.

A connecting rod needs to be checked for bends on all three axes, where many occur. Damaged connecting rods are often twisted, flexed, or stretched.

Twisted

A twisted connecting rod is one where the small end and the big end of the rod are skewed with respect to one another. Obviously the big end of the rod, connected to the crankshaft, will remain on the proper plane. The small end is connected to the piston. When the piston is pushed up the cylinder on the compression stroke, it will be pushed up with the force off center and angular, causing uneven wear. When the piston is forced down by the combustion gasses, the force on the thrust side of the piston will also be angular and off center. The piston is designed to receive forces in a very specific way. When forces are received angular and off center, it will cause stresses in and on the piston that can cause poor wear patterns and may crack or damage the piston.

Flexed

Imagine that you have borrowed your brother's umbrella and have taken it to the racetrack. Due to a lack of grandstand seating, you find that you will have to stand for the duration of the race. Sitting in the grass is out of the question as the ground is muddy. Remembering the umbrella you borrowed, you stick the point in the ground and sit on the handle. All goes well until you shift your weight slightly, and the handle forms a crease in an inappropriate place and direction.

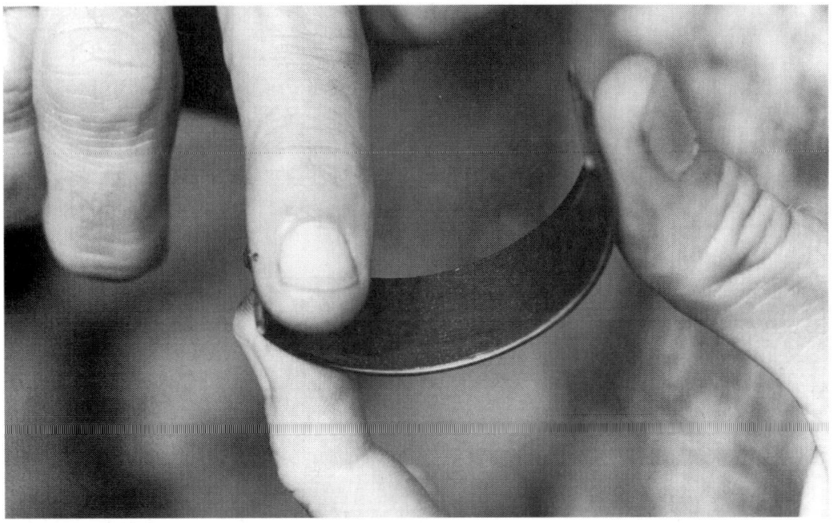

The middle of the shaft of the umbrella shifts to the right while the point in the ground and the handle in the crease remain in their relatively original position. The umbrella is bent. This can happen to a connecting rod for many reasons, high-combustion pressures or coolant in the combustion chamber during

Left
When connecting rods are installed, a quality prelube should be used on the bearing surface. This will prevent damage to the rod bearing while cranking before and during initial start-up.

This connecting rod journal of the crankshaft looks fine at first glance. Even when the crankshaft has such minor wear and damage that no major machine work needs to be done, it should still be polished. Even crankshafts that appear to be perfectly smooth have microscopic burrs that can cause damage to the new bearing. Industry specifications dictate that the crank journals should be polished to a smoothness of 5 to 10 micro-inches (0.000005 to 0.000010in). Visual inspection of the crank journals to determine their need to be polished is wholly inadequate. The human eye has an ability to see objects as small as 10 microns. Ten microns is 394 micro-inches (0.000394in). The smallest object you can see, therefore, is forty times larger than the industry standard for smoothness. Several methods are available for checking the smoothness of the crankshaft. The high-tech method is to use a profilometer. This is an electromechanical device with an accuracy sixty-two times more accurate than your fingernail. The bottom line is to have the crankshaft polished with each engine rebuild. Polishing should also be done after the crankshaft has been machined.

the compression stroke being the most likely.

Stretched

A stretched connecting rod is an unlikely event in the typical family sedan. Earlier, we discussed the tremendous forces that are exerted on the piston. Remember that these forces are transferred to the connecting rod. These forces at high speed and high temperatures can actually lengthen the connecting rod. The amount of lengthening that occurs is quite small but can result in unbalanced compression ratio.

And if it is bent?

As with so many other things, the answer to this question depends on the destined use of the engine. Connecting rods can be straightened. For most engine uses where low rpms and small loads are typical, for instance a family sedan, a straightened rod will work just fine. This is one of the projects we had to successfully complete when I was in trade school. The rod was placed in a vise and a special pry bar was slipped over the portion of the rod that needed to be bent. Several attempts at twisting the rod were needed to get it straightened correctly. Does this seem kind of hokey to you? Me, too.

A connecting rod that has been bent has had the integrity of its structure disturbed. Therefore, the rod is weakened. When the rod is bent back, its structural integrity is disturbed once again. Without a doubt, the better policy is to replace bent connecting rods. This will be discussed more a little later.

Heat damage

Remember in the old westerns—back when the good guys wore white hats and Gabby Hayes always said, "They went thataway, sonny," and it was the truth—the protagonist rode up to the local blacksmith shop. The smithy was always pounding a horseshoe on an anvil with a hammer. He would then plunge it into cold water, then heat it up again. If you ever wondered what he was doing, you are about to find out. He was forging the shoe by beating it with a hammer and then heat-treating it with the forge and the cold water. Heat, properly applied, strengthens the metal. Heat, improperly applied, can weaken the metal.

In many cases, the heat-damaged component can be identified by the bluish shading on the surface of the affected metal. The presence of the blue shading confirms the probability of heat damage; however, the lack of the blue shading does not confirm the lack of heat damage. The best defense against a catastrophic failure is to replace any pistons and connecting rods that have been possibly subjected to extremely high temperatures.

Nicks and burrs

Nicks and burrs are flaws on a component that provide a starting point for cracks and breaks. Torsional and tensile loads that are normally spread out over a large area become concentrated at the flaw.

Inspecting connecting rods for nicks and burrs is a job best suited for the anal retentive. Gauge your personality type. I tend to be more oriented toward getting the engine assembled and getting the vehicle on the road. Realizing that this is a weakness, there are two things I will do to compensate. First, I force myself to inspect every component, square centimeter by square centimeter. Second, I find an individual, perhaps not even an expert, who is very detail-oriented. This type of person sees the universe as the sum of its parts. This is exactly what the ideal inspector needs to be. The ideal inspector, however, may spend so much time on the minute details of the job that the engine will never get completed. On the other hand, a personality type, like mine, who puts an emphasis on completing a job has to be deliberate about inspection of the parts.

While inspecting for nicks and burrs, look closely at the casting seams of the component. Residual flashing can concentrate stresses

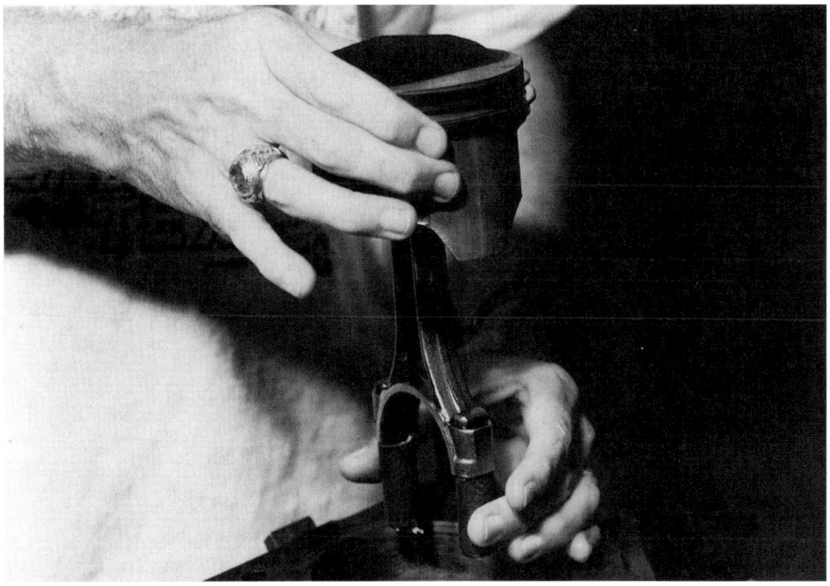

Long pieces of rubber hose can help guide the connecting rod into its proper position on the crankshaft. Gently lower the piston into position.

When installing the bearing caps, be aware that there is only one way the bearing is supposed to be installed. The caps and connecting rod are often marked, as in the photo above. If the rod and cap are not marked, then make sure that the bearing tabs are both on the same side of the crankshaft journal.

the same as other minor defects. When these flaws are found, they should be filed smooth. Obviously there may be a judgment call required on your part. If removing the defect will weaken the component more than the burr, flashing, or nick, then the part should simply be replaced.

Residual magnetism

Residual magnetism is a problem that is often not considered, even by the most experienced of engine builders. Iron and steel components can become magnetized as a result of the hammering caused by excessive clearances before the engine was disassembled or as a result of machine work. Magnetic inspections, magnafluxing, and the like can also magnetize components. When a component is magnetized, it attracts all the little bits and flakes of metal that result from normal engine wear. These bits and flakes belong in the bottom of the oil pan attached to that magnetic oil drain plug, not attached to expensive engine parts.

A magnetized part can be identified by laying a thin sheet of paper over the part and sprinkling iron filings or dust on the sheet of paper. If the iron filings arrange themselves in distinct patterns, the part is magnetized. To demagnetize a component, borrow a trick from the television repairman. In the TV repair industry there is a tool called a degausser. This tool creates a powerful magnetic field designed to scramble the magnetic fields of other objects. A similar, but less powerful, device used to erase audio and video tapes can be purchased from stores like Radio Shack. During my brief career as a television/electronics technician, I was taught to use a powerful soldering gun to degauss picture tubes. Much to my surprise, it also works on magnetized engine parts.

Whichever tool you have access to or choose to use, the technique is the same. Energize the degaussing device away from the component to be demagnetized. Pass the degausser back and forth across the components in random patterns. Make sure each pass stops well beyond the component. After degaussing the part, check for magnetism again using the iron filing technique described above.

Note: Be careful not to allow any iron filings to get on the component. If they do, make sure that they are thoroughly cleaned off before installing the component.

Big end and small end alignment

A twisted connecting rod can occur as a result of a wide range of events. Earlier I told the story of the submerged Volvo; that was one way. Other causes of twisting include detonation, blown head gaskets, foreign objects, broken valves, broken timing chains or belts, and overrevving.

To test for a twisted connecting rod, a precision gauge is required. This gauge costs several hundred dollars, and it is unlikely that you will want to purchase one for just one engine. On the other hand, if you are going to be rebuilding engines or this engine several times, then it may be worth the investment. Now keep in mind as you continue to read this paragraph that within my personality as a technician is a driving desire to do expensive procedures in an inexpensive manner. There are several tricks I have learned over the years to check the alignment of connecting rods. None of these tricks is very effective. The bottom line is that in this case especially, the right tool is the only tool. If you suspect that the connecting rods are bent or twisted, take them to a machine shop that is prepared to straighten them, or replace them.

Big end and small end eccentricity

Several times already in this book you have read about the effects of inertia on the piston and connecting rod assembly.

After the pistons are installed and the connecting rods are torqued, check the connecting rod side clearance for each rod.

My dog has a tug-of-war toy that is a lot like a connecting rod. It consists of a rubber "shaft" connected to round loops of rubber. Often he will bring this toy to me when he wants to play tug-of-war. He will bite one loop in his teeth and I will hold the loop on the opposite end. As we pull in opposite directions, the loop on each end will elongate. This is what can happen to connecting rods with years of use and seconds of abuse. Since the end connected to the crankshaft is larger, there is more of a tendency for this big end to elongate. Use an inside micrometer or telescoping gauge to determine if either end of the connecting rod is out of round. Measure the bore of the big end on both the X and Y axes. These measurements should be within 0.001in of one another.

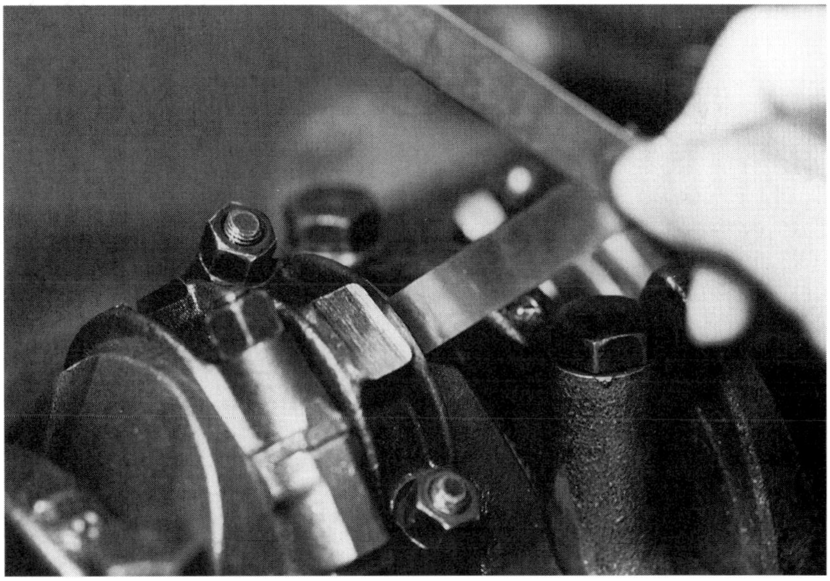

Hold the connecting rod as far to one side as possible. Measure the side clearance using a feeler gauge. Verify the acceptable range of clearances with the specifications. Typical side clearances are 0.010 to 0.020in. If the side clearance is outside the specified range for this application, replace the connecting rod. If the side clearance is still incorrect, there is a problem with the crankshaft.

Repairing (rebuilding) a connecting rod

If the big end of the connecting rod does not meet the specs for eccentricity or if the connecting rod is bent, it can sometimes be repaired. When considering the repair of a connecting rod, also consider how the engine for which the connecting rod is destined will be used. Performance engines and engines used in wilderness areas should have new connecting rods to replace those that are damaged. A few hundred dollars saved on not replacing questionable connecting rods is not worth your life.

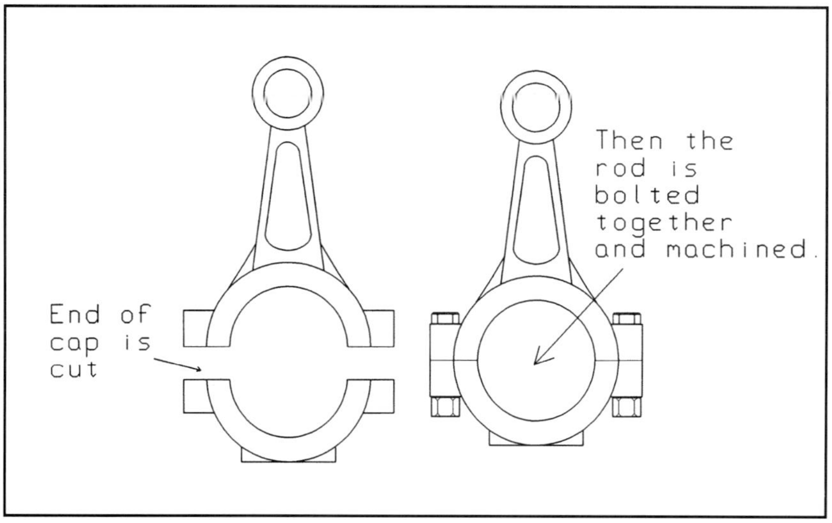

If the big end of the connecting rod is damaged or elongated, the ends of the caps are filed shorter. Then the cap is bolted in place and the big end hole is machined back to spec.

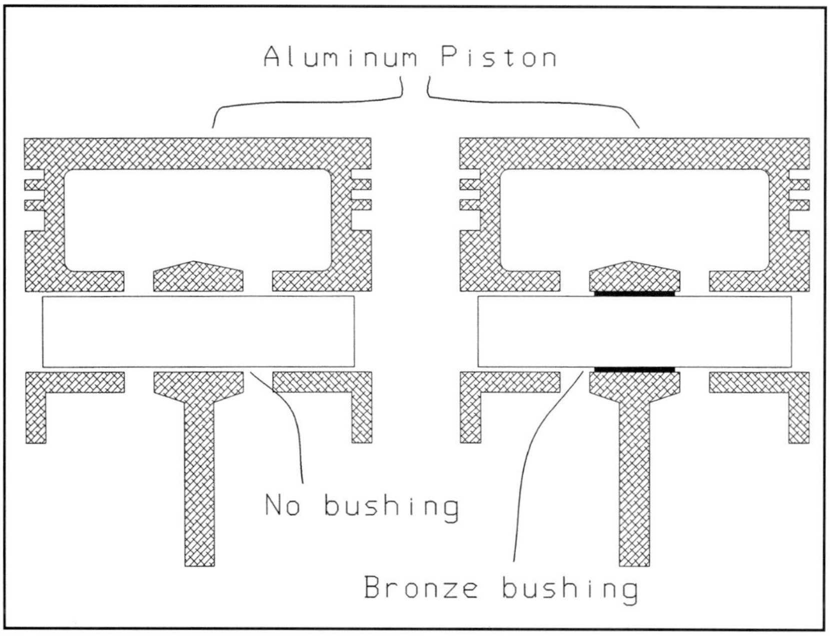

If the small end of the connecting rod is damaged, it can be machined larger. A bushing is then pressed into place. Some connecting rods, especially those that are used in applications where the wrist pin is pressed into the piston, have bronze bushings in the small end as a standard feature.

Connecting rods that have a minor bend or twist in them can be corrected by simply reversing the bending process. Pad the jaws of a vise with soft wood and clamp the rod in the vise. The wood will protect the rod from the teeth of the vise. With the piston mounted on the connecting rod, slide a long, thin pry bar into the wrist pin. Gently apply pressure to the connecting rod in the direction that you wish it to straighten. Gradually increase the pressure. You may feel the rod bend or you may not. Try to bend the rod only a little at a time. Each time you bend the rod it is weakened. You do not want to bend the rod back and forth so many times that it weakens; remember that as you rock the pull tab of a beer can back and forth, it eventually weakens to the point where it breaks off.

Warning: do not attempt to straighten the connecting rod without the piston and wrist pin installed, as severe damage to the connecting rod may result.

Connecting rod ends that have been stretched eccentric can be repaired by most machine shops. The procedure begins by machining the mating surfaces of the connecting rod cap and the connecting rod. This will make the average diameter of the big end smaller, but will not remove the eccentricity. The big end bore is now machined back to the original truly round diameter.

All this talk about rebuilding connecting rods belies the fact that if an engine is being built for performance or durability, new rods should be used. This is especially true if there is any question about the structural integrity of the old rods.

Balancing

Frankly, the best way, and often cheapest way, for the individual interested in street performance to balance his or her connecting rods is to buy them that way. However, if the challenge is more important than common sense, and for a true motorhead that is the way it should be, or if this engine is truly destined for the racetrack, then balancing skills are required. The key to balancing an engine is to ensure, as much as is possible, that each set of reciprocating masses is as equal as possible.

The first and most essential piece of equipment for balancing connecting rods is a good set of scales. These scales should be accurate to about half a gram. That's about half the weight of a dollar bill. It may seem obvious to some, but begin by weighing all the connecting rods. Use the lightest rod as the standard. At each end of the connecting rod there is a flat area. You may have thought that the purpose of this flat area was to allow you to stand the connecting rod on end. Actually, they are called balancing pads. To balance the connecting rods, grind or file small amounts of metal from these pads. Keep in mind that the heat generated while grinding the balance pads can change the nature of the metal, usually weakening it. Filing, although much slower, generates far less heat and therefore is far less likely to do damage to the connecting rod. If you choose to grind, as I confess I often do, limit the time the connecting rod is on the grinding wheel. Limiting the time will limit the risk of heat damaging the rod.

When all the connecting rods weigh the same to within half a gram, you are ready to install the pistons.

Earlier we discussed the balancing of the pistons. Logically, if the balanced pistons are installed on balanced connecting rods, the reciprocating masses will be balanced, right? If an honest man is elected to Congress, he will remain honest, right? Weigh the reciprocating assemblies, including bearings, to confirm balance. If you find that they are not balanced, fine tune the balance before assembly.

If your goal is to create a rapidly accelerating racing engine, then lightening the connecting rods can reduce the reciprocating mass and increase performance. However, even if the goal is simply to give mom and pop a smooth trip to the local Tastee Freeze, balancing the rods is essential. Like the pistons discussed earlier, the connecting rods should be within 1/2 gram of one another. In fact, the entire reciprocating mass, connecting rod, piston, rings, and wrist pin should be balanced to within 1/2 gram of one another as a unit.

Installation

Offset

Several years ago, I was teaching a class about automotive electronics in Canada. When I teach, I often use examples of professional mistakes I have made. Many of these stories you have read in this book or in my other Motorbooks International publications already. After several weeks of these stories, one of the students commented, "You must be a terrible mechanic." Okay, here is another one. Remember the submerged Volvo 265 mentioned earlier in this book? After completing an overhaul of this engine, it had a rapping noise coming from the

This piston exhibits an effort to maximize compression and squeeze every possible ounce of power out of the engine. The piston protrudes about 0.025in above the deck of the block.

middle portion of the block near cylinder number one. Henry, a rather round and highly experienced mechanic whose basic philosophy of life was that work interfered with fishing, said, "Sounds like a piece of carbon in the number one cylinder." When I reminded him that I had just rebuilt the engine, he said, "Well, I guess you should have cleaned the heads and intake better." He recommended that I just drive the car. He said that it would soon pass like a kidney stone. I decided to tear the engine apart again. After careful inspection, I determined that I had installed the connecting rod on the number one piston backwards. When the engine was not running, there was plenty of

The builder chose a 0.050in head gasket. This, in theory, should provide a clearance of 0.025in when the head is bolted into place.

clearance between the moving parts; however, when the engine was running, combustion forces on the top of the piston shortened the connecting rod just enough to cause the skirt of the piston to tap lightly on the head of the rod bolt. The rods in this engine had enough offset that the skirt would tap when the rod was installed backwards.

Most connecting rods have a mark on them to identify the proper direction of installation. These marks are sometimes in the form of a delta or an arrow. Care should be taken during disassembly to identify each of the connecting rods and note the direction of installation. Newly purchased connecting rods often come with installation instructions.

Buying performance connecting rods

Speaking of newly purchased connecting rods, the best way to reduce reciprocating mass, lower energy loss, and increase power is to purchase high-performance aluminum connecting rods. This is especially important when you have purchased new pistons.

At this point, it would be easy for this book to turn into a catalog of the connecting rods available. Rather than provide free advertising for countless companies across the country, let's take a look at the decision-making process. What will this engine be called upon to do?

1. Take mom and pop to the grocery
2. Street performance
3. Roundy-round short track racing
4. Roundy-round duration racing
5. Competition drag racing
6. Maximum power drag racing

Common connecting rod options are:

Cast—These are inexpensive and durable. These connecting rods are the ones normally found in stock engines. With a little work, these rods can perform effectively in a street performance application or a bracket racer.

Forged premium steel—These connecting rods were the mainstay of the performance engine industry for many years. They have the combined advantages of strength, lightness, and durability. This is generally one

This is the top of the 0.030in oversized TRW piston installed in this engine.

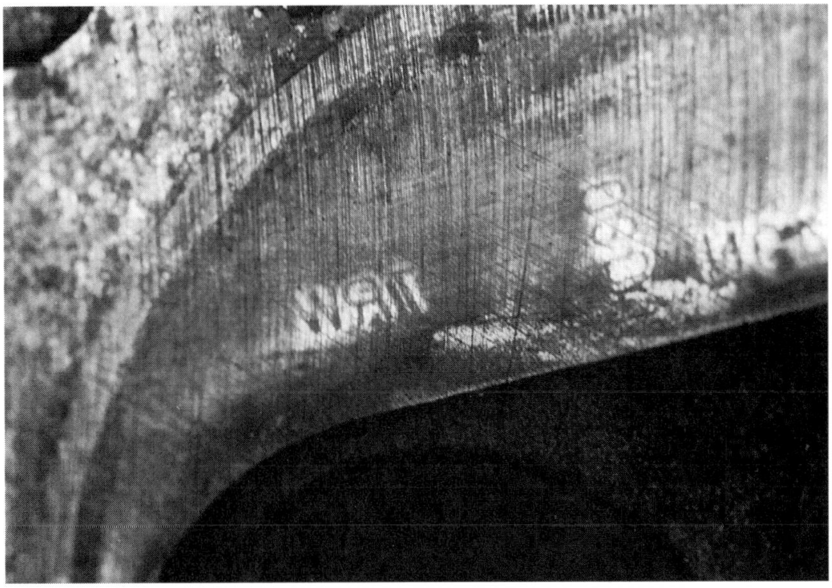

This is the bottom of the head after the engine has been run for a few hours. Notice the 0.030in and the TRW marks are exactly a mirror image of the markings on the top of the piston. Although there was a theoretical 0.025in clearance, the engine builder failed to keep in mind that when the piston reaches top dead center, the inertial force of the piston tries to keep the piston moving toward the cylinder head. Consequently, the connecting rod stretches slightly. In this case, the amount of stretch was just enough to allow the head of the piston to just barely touch the head in the connecting rod. The engine builder should have chosen a rod with less stretch or provided a little less positive deck height.

of the best rods to be used in street performance machines.

Cast aluminum—Aluminum connecting rods are light and strong. Their weakness comes from low fatigue strength. This means low durability. Frequent replacement of aluminum connecting rods is necessary. Aluminum connecting rods should not be used in engines where durability is of the essence. Endurance racers, aircraft, and family cruisers should avoid the use of aluminum.

Performance preparation of a stock connecting rod

Magnafluxing

Remember when your father asked you why your "Sergeant Pepper's" poster was painted in such strange colors? Remember how you would wait for him to leave the room then turn that strange special light back on? That light was a black light. If you have fought for years with your spousal unit about whether or not to throw that "silly thing" out, you are now reprieved. You no longer have to tell her or him, "I'm saving it in case Jimmy Hendrix makes a comeback."

Magnafluxing is a procedure that uses a large electromagnet to magnetize each connecting rod. A special magnetic fluid is then saturated into the rod. When examined under a black light, surface cracks will be revealed. These surface cracks can be the starting point of much more serious cracks and breaks.

Note: After performing a magnafluxing procedure, be sure to demagnetize the rod using the procedure addressed earlier in this chapter. Leaving it magnetized will allow it to attract steel and iron particles after the engine is started.

X-raying

If you are blessed as one of those people who has more money than you do spousal and progenous demands for the money, then x-raying the connecting rods will reveal not only surface cracks, but cracks below the surface. This is an expensive procedure and a procedure that is not even available in many towns, but it is well worth the expense for peace of mind. On the other hand, if you are really in doubt about the connecting rod, in doubt enough to consider the expense of having the connecting rods x-rayed, save the money on the x-ray and replace the rods.

Polishing

Polishing is done to strengthen the connecting rod, not to make it look nice. Some rules in stock racing classes do not permit polishing. If the engine you are building is not bound by rules or if the rules do not prohibit polishing, it will be time well spent. It will also be cheaper than spending $1,000 on a set of performance connecting rods.

Use a high-speed die grinder and a carbide cutter to remove the casting or forging marks from the sides of the connecting rods. This operation is often referred to as deburring. With a stationary belt sander, grind the casting pits and marks off of the connecting rods. This operation denies cracks a place to start, making the rod more durable. Smoothing the rod beams removes the small pits and crevices that provide a starting point for cracks. The rods are instantly stronger.

Shot peening

After polishing, the connecting rods can be further strengthened through shot peening. After masking the bearing mating surfaces to protect them, high-velocity shot is targeted at the other areas of the connecting rods. Particular attention is paid to the fillet or radius near the rod bolts. The impact of the shot relieves stress in the metal that could cause microscopic cracks that could later become a broken connecting rod. The principle is not unlike stopping at the massage booth at the Seattle airport to have your back pounded on to relieve stress. Additionally, shot peening compacts the surface of the rod, providing a homogeneous surface that allows no place for cracks to begin.

Bushing

A lot of attention has been paid thus far to the big end of the connecting rod. The small end can also be subjected to damage and require repair. Since the small end of the rod cannot be separated like the big end, repairing the small end is not as easy as repairing the big end. The small end must first be machined round. Then a bushing is driven in place and flushed off with the side of the rod.

New bolts

Installing new connecting rods or rebuilt connecting rods without using new rod bolts is like splicing an anchor chain with a piece of twine. Although it will initially serve the function, it may prove to doom the life of the work that has been done by providing a weak point for failure.

Chapter 5

Crankshafts

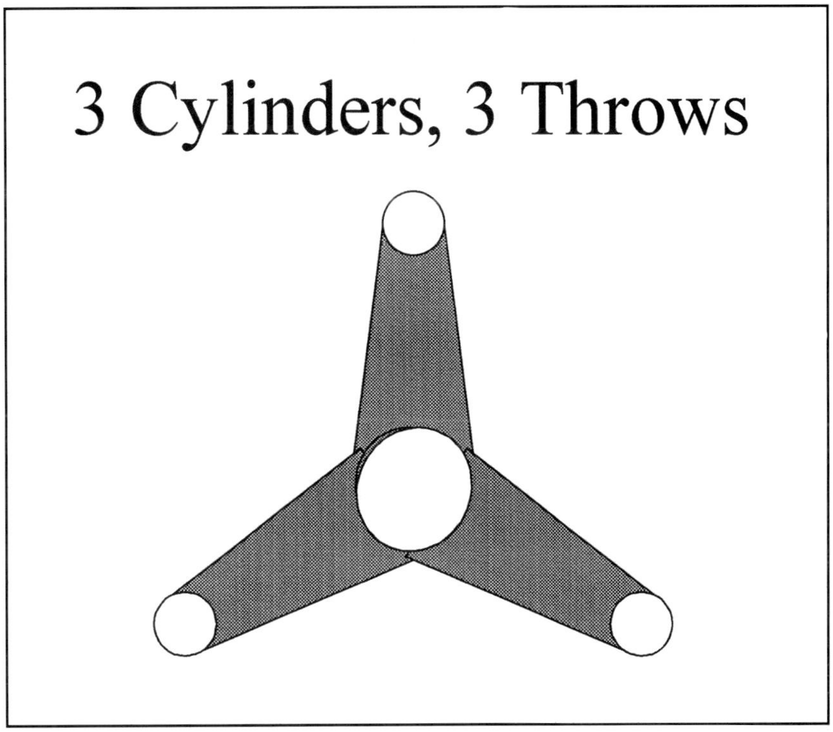

The three-cylinder engine configuration locates the connecting rod journals at 120 degree intervals. Therefore, the cylinders fire at 240 degree intervals.

Crankshaft configurations

The power created by the engine is conveyed to the transmission through the crankshaft. It could be said that the crankshaft converts the vertical action of the pistons in the cylinder to the rotational action required to move the car down the road.

The crankshaft is either forged or cast from a variety of alloys. While the forged crankshaft is usually stronger, it is also expensive. The cast crank is more than adequate for most nonracing uses.

After casting or forging, the bearing surfaces are machined and are then fine ground and polished. Modern crankshafts feature counterweights that balance it against the weight of the pistons and connecting rods. Further balancing is accomplished by shaving and drilling the counterweights.

Dynamics of the crankshaft

What the crankshaft does is actually quite phenomenal. During combustion, extreme gas forces are placed on a single connecting journal of the crankshaft forcing it to accelerate. Meanwhile, depending on engine design, other connecting rod journals are experiencing acceleration, deceleration, neutral loads, and negative loads. Throughout all of this the crankshaft is converting reciprocating motion into rotary motion. Throughout the conversion process the crankshaft must make this conversion smoothly and with a maximum of comfort.

Crankshaft configurations

There is nothing haphazard about the arrangement of the journals on the crankshaft and there should be nothing haphazard in choosing the platform for your powerplant. There are reasons that certain engine configurations are more popular for performance engines and for luxury car engines than others. We will now take an in-depth look at why an opposed six-cylinder has advantages over a 60 degree V-6 and why an in-line six-cylinder has some advantages over a V-8.

Three-Cylinder in-line (three throws)

A three-cylinder engine? Yes, there have been three-cylinder engines on applications other than motorcycles. One of my favorite little engines of all times is the three-cylinder SAAB engine of the late 1960s. This was a two-stroke three-cylinder engine.

This engine configuration locates the connecting rod journals at 120 degree intervals. The cylinders, therefore, fire at 240 degree intervals. The vertical forces of the combustion process on the crankshaft are negligible. The tangential forces of combustion are also negligible. Vertical and tangential inertial forces are, however, considerable. End result: the engine tends to shake a little when the engine is running.

Four-Cylinder in-line (four throws)

This crankshaft design locates the connecting rod journals

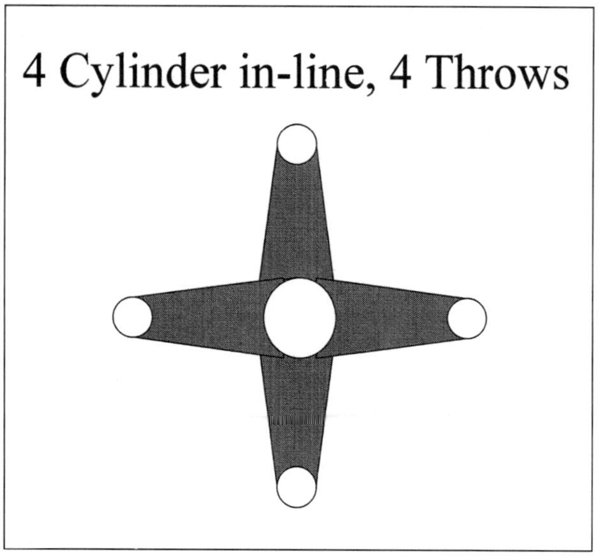

This crankshaft design locates the connecting rod journals at 180 degree intervals along the crankshaft. Cylinders one and four are on one side of the centerline of the crankshaft. The journals for cylinders two and three are located at 180 degrees of one and four on the opposite side of the centerline of the crankshaft.

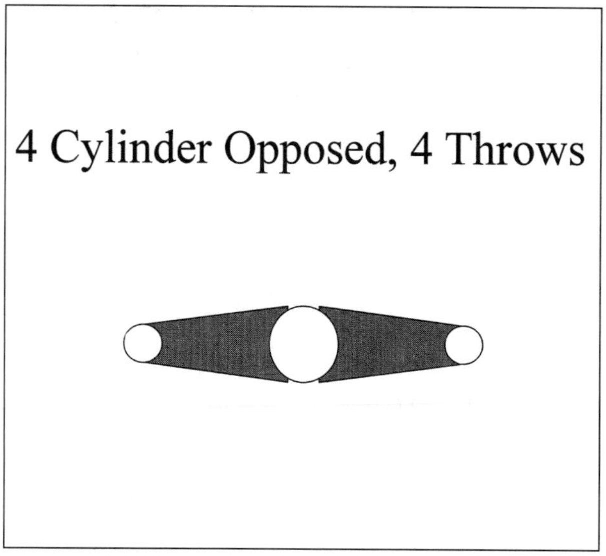

This configuration excels in low vertical and tangential combustion forces. Like the in-line four-cylinder, the connecting rods for cylinders one and four are on one side of the centerline of the crankshaft. The journals for cylinders two and three are located at 180 degrees of one and four on the opposite side of the centerline of the crankshaft.

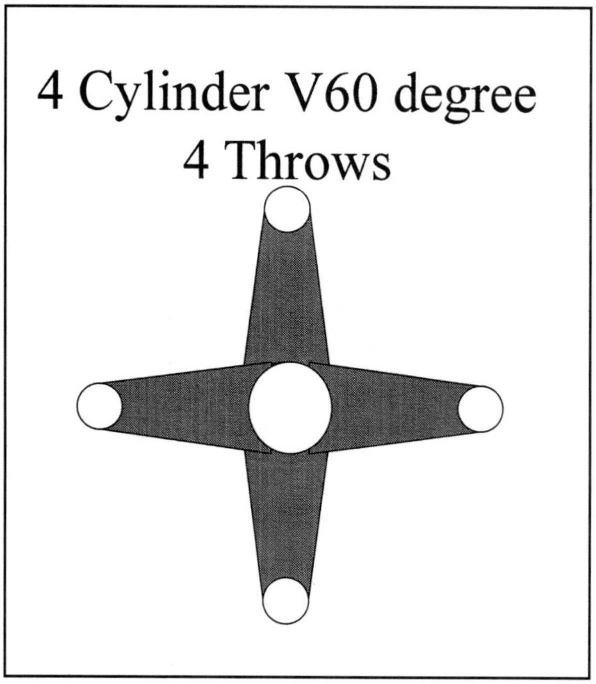

There have been a few applications during the history of the automobile that have featured a V-4 engine. This configuration features four rod journals along the length of the crankshaft. Imagine the first journal, cylinder number one, to be located at zero degrees. In a clockwise direction, the second journal is at 270 degrees, the third at 180, and the fourth at 90 degrees.

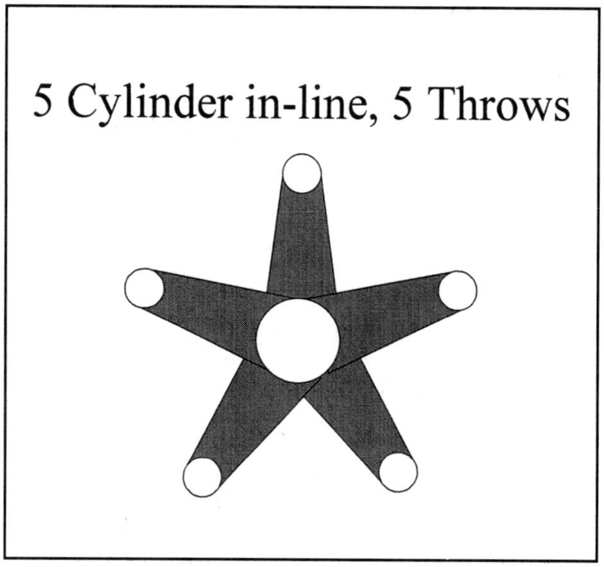

There once was a commercial where the narrator, using a thick, phony German accent said, "They asked us why five cylinders in the Audi 5000? We said because four was not enough and six were too many." As far as I know, in the world of gasoline-fueled, spark-ignition engines, the Audi 5000 was the only one to use an in-line five-cylinder engine. The second cylinder in the firing order is cylinder number two. The journal for cylinder number two is located 144 degrees clockwise from the journal for cylinder number one. The next in the firing order is cylinder number four, located 288 degrees from cylinder one. Then number five, which is 72 degrees from number one. The last cylinder to fire is number three, which is 216 degrees from number one. The cylinders fire 144 degrees apart.

at 180 degree intervals along the crankshaft. Cylinders one and four are on one side of the centerline of the crankshaft. The journals for cylinders two and three are located at 180 degrees of one and four on the opposite side of the centerline of the crankshaft.

The in-line four-cylinder engine is free of undesirable forces of inertia. The vertical forces of combustion are also extremely limited. However, the tangential forces of combustion will cause the engine to rock laterally when it is running.

Four-Cylinder opposed (four throws)

This configuration excels in low vertical and tangential combustion forces. Like the in-line four-cylinder, the connecting rods for cylinders one and four are on one side of the centerline of the crankshaft. The journals for cylinders two and three are located at 180 degrees of one and four on the opposite side of the centerline of the crankshaft.

The opposed four-cylinder configuration has low vibration effects from vertical and tangential forces of combustion. Additionally, vertical forces of inertia are virtually nonexistent. Tangential forces of inertia are relatively high; the engine will tend to rock as it runs.

Four-cylinder V-60 degree (four throws)

There have been a few applications during the history of the automobile that have featured a V-4 engine. This configuration featured four rod journals along the length of the crankshaft. Imagine the first journal, cylinder number one, to be located at zero degrees. In a clockwise direction, the second journal is at 270 degrees, the third at 180 degrees, and the fourth at 90 degrees.

These engines have low vertical forces from combustion, but all other forces are very high. This engine configuration vibrates a great deal.

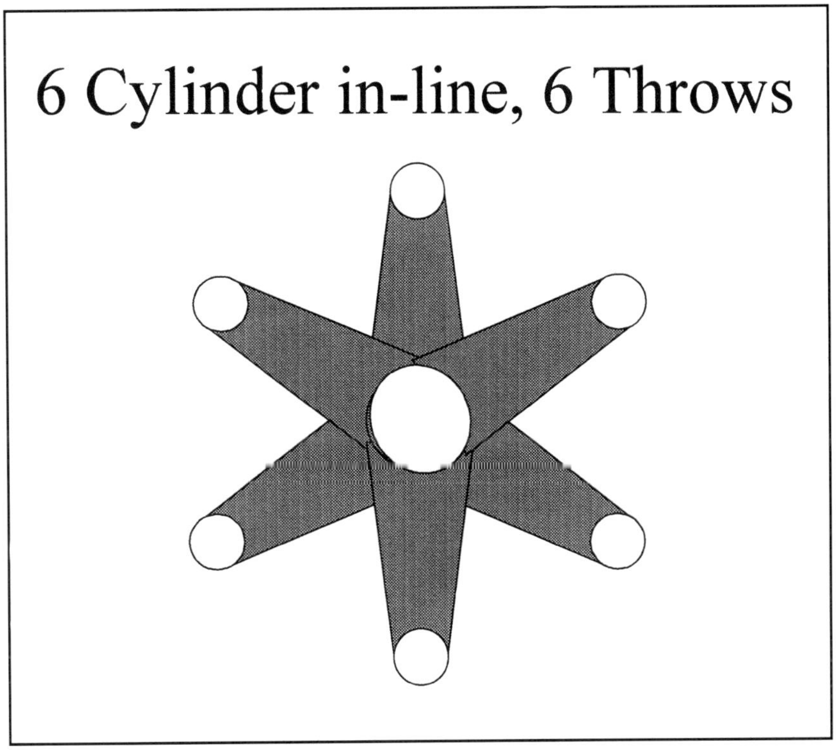

This design is a classic. It has been used by almost every automobile manufacturer in the world. Anytime there is a system that is so widely used, there must be a good reason for it, and there is.

Five-cylinder in-line (five throws)

There once was a commercial where the narrator, using a thick, phony German accent said, "They asked us why five cylinders in the Audi 5000? We said because four was not enough and six were too many." As far as I know, in the world of gasoline-fueled spark-ignition engines, the Audi 5000 was the only one to use an in-line five-cylinder engine. The second cylinder in the firing order is cylinder number two. The journal for cylinder number two is located 144 degrees clockwise from the journal for cylinder number one. The next in the firing order is cylinder number four, located 288 degrees from cylinder one. Then number five, which is 72 degrees from number one. The last cylinder to fire is number three, which is 216 degrees from number one. The cylinders fire 144 degrees apart.

In the five-cylinder in-line design, vibrations resulting from combustion forces are negligible. Inertial forces, both vertical and tangential, do result in vibration, but this vibration is minimal. All in all, as odd as five cylinders sounds, it is a pretty credible configuration for an engine.

Six-cylinder in-line (six throws)

This design is a classic. It has been used by almost every automobile manufacturer in the world. Anytime there is a system that is so widely used, there must be a good reason for it. There is. There are several variations on this engine configuration. Generally speaking, it places the rod journal for the second cylinder in the firing

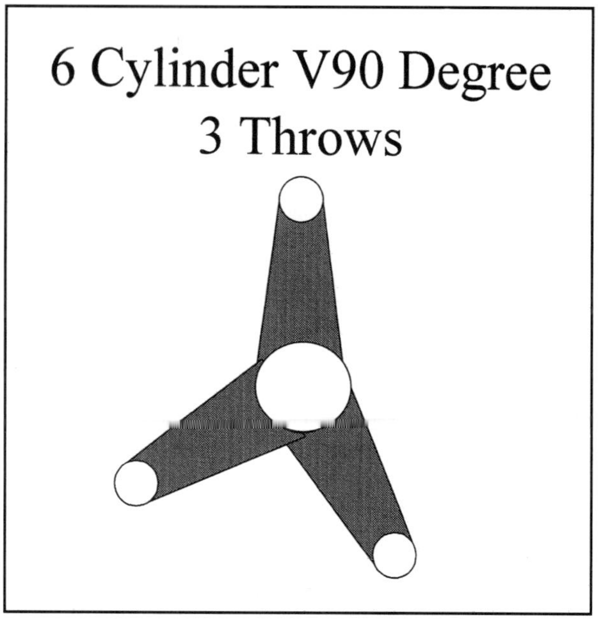

This is a weird setup. There have been several 90 degree V-6 engines built worldwide. The following comment is intended to be descriptive rather than statistically accurate. It seems that the reputations of the 90 degree V-6 engine that have been produced have taken on a life of their own.

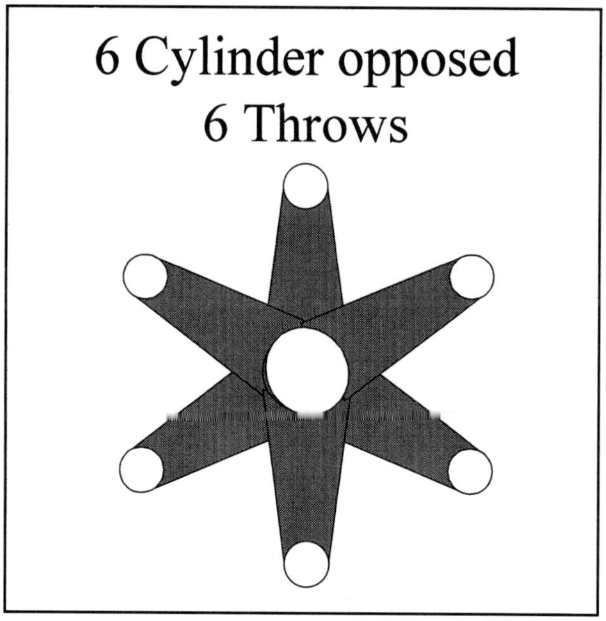

This engine places the rod journal for the second cylinder in the firing order either at 120 or 240 degrees from the firing of the first cylinder. The third cylinder of the firing order will be at 240 or 120 degrees from number one in the firing order. The number four journal will be 0 degrees (or, if you prefer, 360 degrees) from number one. Journal five will be at 240 or 120 degrees, and the sixth journal will be at whatever angle the fifth one is not. The cylinders fire at 120 degree intervals.

order either at 120 or 240 degrees from the firing of the first cylinder. The third cylinder of the firing order will be at 240 or 120 degrees from number one in the firing order. The number four journal in the firing order will be 0 degrees (or if you prefer 360 degrees) from number one. Journal five will be at 240 or 120 degrees, and the sixth journal will be at whatever angle the fifth one is not at. The cylinders fire at 120 degree intervals.

This engine is a near-perfect design from a vibration and stress perspective because all forces are balanced. In fact, the crankcase volume is even a constant. As pistons number one and six drop on their combustion and intake strokes, thereby potentially decreasing the volume of the crankcase, the other pistons are moving to increase the volume.

Six-cylinder V-90 degree (three throws)

This is a weird setup. There have been several 90 degree V-6 engines built worldwide. The following comment is intended to be descriptive rather than statistically accurate. It seems that the reputations of the 90 degree V-6 engine that have been produced have taken on a life of their own. The crankshaft of this engine has three throws. To describe this engine we must talk about the full 720 degrees of crankshaft rotation associated with a complete engine cycle. From cylinder number one, the number two cylinder's connecting rod journal will be located 150 degrees in a clockwise direction. The third connecting rod's journal will be 240 degrees from number one. The fourth cylinder will be 390 degrees from number one, the fifth 480 degrees from number one, and the sixth 630 degrees from number one. This leaves 90 degrees of rotation remaining before number one is to fire again.

Now you may be thinking at this point, "That is a rather odd way to fire an engine." There would be many who would agree with you. This engine is often referred to as an odd-fire configuration.

Vertical and tangential vibrations due to combustion are minimal. However, the vibrations caused by inertia are considerable. This engine will both bounce and rock.

Six-cylinder opposed (six throws)

This design is also a classic. It has been used by almost every aircraft engine manufacturer in the world. Generally speaking, this engine is simply a pancaked version of the in-line six-cylinder. It places the rod journal for

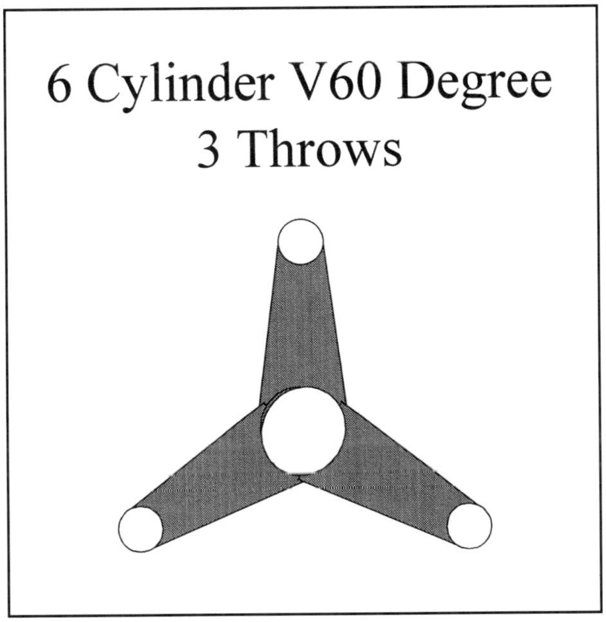

Unlike the 90 degree V-6, this is an even-fire engine. Like the 90 degree V-6, this engine design also has vertical and tangential vibrations from combustion that are minimal and vibrations caused by inertia that are considerable. This engine will both bounce and rock.

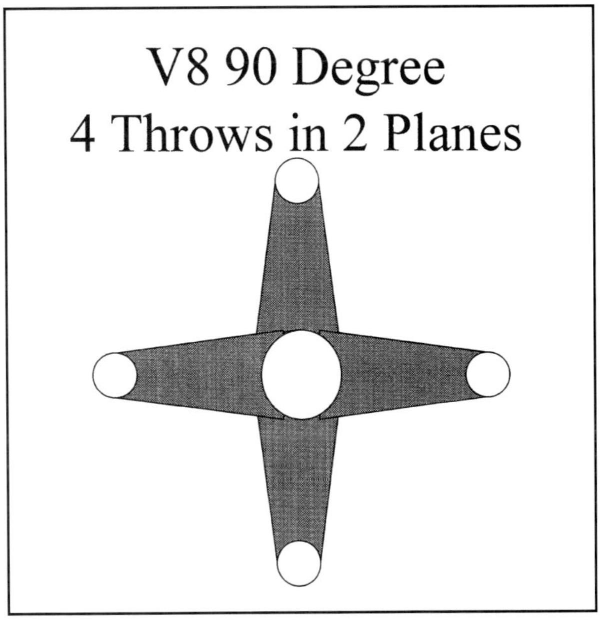

For several decades now the V-8 engine has been the mainstay of the "big iron" automotive industry and the racing industry. Again, the engine design must be successful at solving some fundamental problems to warrant this adulation. The design features four connecting rod throws in two planes. Each throw is 90 degrees apart.

the second cylinder in the firing order either at 120 or 240 degrees from the firing of the first cylinder. The third cylinder of the firing order will be accordingly at 240 or 120 degrees from number one in the firing order. The number four journal in the firing order will be 0 degrees (or, if you prefer, 360 degrees) from number one. Journal five will be at 240 degrees or 120 degrees, and the sixth journal will be at whatever angle the fifth one is not. The cylinders fire at 120 degree intervals.

Like its in-line cousin, this engine is a near-perfect design from a vibration and stress perspective because all forces are balanced. In fact, the crankcase volume is even a constant. As pistons number one and two drop on their combustion and intake strokes, thereby potentially decreasing the volume of the crankcase, the other pistons are moving to increase the volume.

Six-cylinder V-60 degree (three throws)

This is the engine design used frequently in front-wheel drive applications. The connecting rod journals are located in three planes 120 degrees apart. Unlike the 90 degree V-6, this is an even-fire engine. Like the 90 degree V-6, this engine design also has vertical and tangential vibrations from combustion that are minimal, and vibrations caused by inertia are considerable. This engine will both bounce and rock.

Eight-cylinder V-90 degree (four throws in two planes)

For several decades now the V-8 engine has been the mainstay of the "big iron" automotive industry and the racing industry. Again, the engine design must be successful at solving some fundamental problems to warrant this adulation. The design features four connecting rod throws in two planes. Each throw is 90 degrees apart. This means that the first cylinder will fire, then the second in the firing order will fire at 90 degrees of engine cycle rotation, the third at 180 degrees, the fourth at 270 degrees, the fifth at 360 degrees, the sixth at 450 degrees, the seventh at 540 degrees, and the eighth at 630 degrees. This leaves 90 degrees back to number one. Since the rod journal requirements for proper firing are 90 degrees apart, this means that pairs of connecting rods can "share" a common journal. Sharing these journals allows for compactness of design. To put it directly, the greatest possible power with the smoothest possible operation in the smallest possible package.

Though not as perfect as the in-line or pancake six-cylinder, this engine design's only vibration shortcoming is vertical vibration from inertial forces. Yet, even this vibration is minimized.

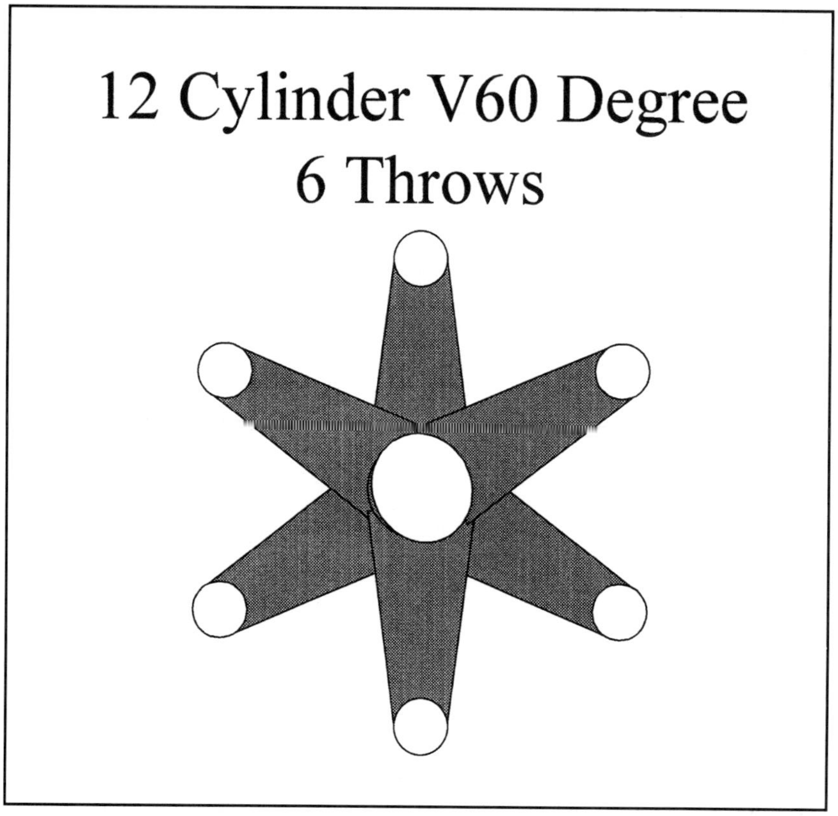

The ultimate engine design, except in terms of compactness, is the V-12 engine. There are many so-called exotic automobile manufacturers that use the V-12 design. In many ways this design is like gluing two in-line six-cylinder engines together and allowing them to share a common crankshaft. Cylinders fire evenly at 60 degrees apart.

Twelve-cylinder V-60 degree (six throws)

The ultimate engine design, except in terms of compactness, is the V-12 engine. There are many so-called exotic automobile manufacturers that use the V-12 design. In many ways this design is like gluing two in-line six-cylinder engines together and allowing them to share a common crankshaft. Cylinders fire evenly at 60 degrees apart. The design is handicapped by neither the combustion force vibration nor the inertial force vibrations.

Stroke

Stroke is the distance that the connecting rod journal of the crankshaft will cause the piston to move in one-half rotation of the crankshaft. Determining the most desirable stroke for an engine is not an easy proposition. A long stroke can increase the leverage of the crankshaft and therefore the torque potential of the engine. A long stroke, however, means a long connecting rod. A long connecting rod means more potential for bending, breaking, and twisting. A short stroke means less potential torque but greater resistance to bending, breaking, and twisting. Long stroke engines inherently have a lower maximum rpm, while short stroke engines are inherently capable of higher rpms.

Vibration

Energy conservation

All of this talk about vibration, tangential forces, and vertical forces actually does have a purpose. Vibration is energy that is being wasted. As you have seen in the discussion of engine and crankshaft configurations,

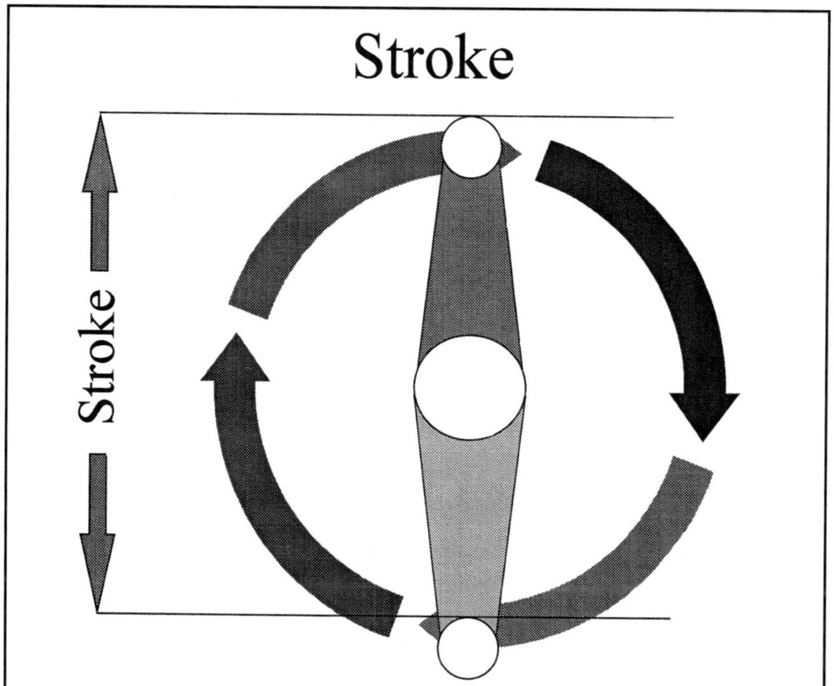

Left
Stroke is the distance between the top side of the rod journal with that journal in the top dead center position and the top of the same rod journal when it is at bottom dead center.

the engine designer takes great pains in weighing package size against power and vibration requirements. When you are building an engine, especially a performance engine, the less vibration you end up with the more power you will have to the ground.

Power loss

In addition to vibration, there are several other factors that rob power from the engine. The internal-combustion, spark-ignition, gasoline-fueled engine is not a very efficient beast, even when everything is working as perfectly as possible. About 30 percent of the potential power from the burned fuel is lost as heat to the engine coolant. About 33 percent of potential power is lost to exhaust heat. About 3 percent is lost to the operation of the cooling fan and water pump, 3 percent to friction, and 4 percent to the pumping action of inhaling air and exhaling exhaust gases. This leaves only about 27 percent power or potential power available to leave the crankshaft.

Back when Burt Reynolds had his own hair, we used to think that raw power from an engine meant that it ran rough. The reality of course is that the smoother the engine runs, the more power it is putting out the back of the crankshaft and the more power it is sending toward the ground. This does not mean to imply that the engine must be smooth under all operating conditions, but rather that it is smooth at the speed the engine is normally operated at. As you probably know, many racing engines idle very roughly, but smooth out at high rpms.

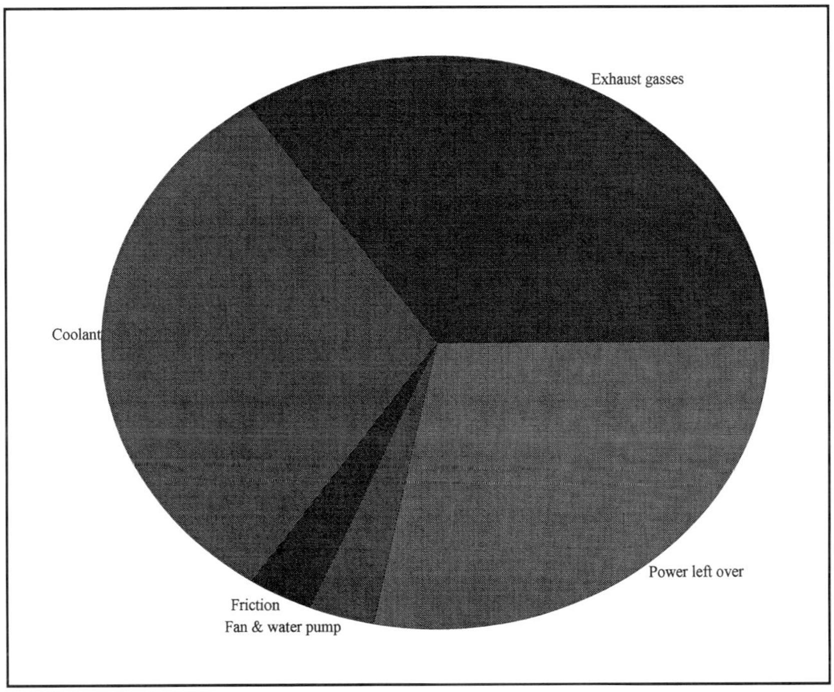

About 30 percent of the potential power from the burned fuel is lost as heat to the engine coolant. About 33 percent of potential power is lost to exhaust heat. About 3 percent is lost to the operation of the cooling fan and water pump, 3 percent to friction, and 4 percent to the pumping action of inhaling air and exhaling exhaust gases. This leaves only about 27 percent power or potential power available to leave the crankshaft.

Types and metallurgy

Cast iron

For reasons of economy and adequacy, the cast iron crankshaft has been the mainstay of the automotive industry for decades. While it may not have the glamour of forged steel or a billet crank, cast iron crankshafts are usually more than adequate for anything licensed for use on the road. Their strength can be further enhanced by some of the strengthening procedures discussed later in this chapter.

There are two primary shortcomings with the cast iron crank. First, the granular structure of the crankshaft is loose and randomized. A lack of structural organization and integrity lessens the potential strength of the component. Second, there is a distinct parting line where the casting halves meet. This not only looks "uncool," it also provides a starting point for cracks and breaks in the crankshaft.

Forged steel

The forged steel crankshaft sports a much denser molecular

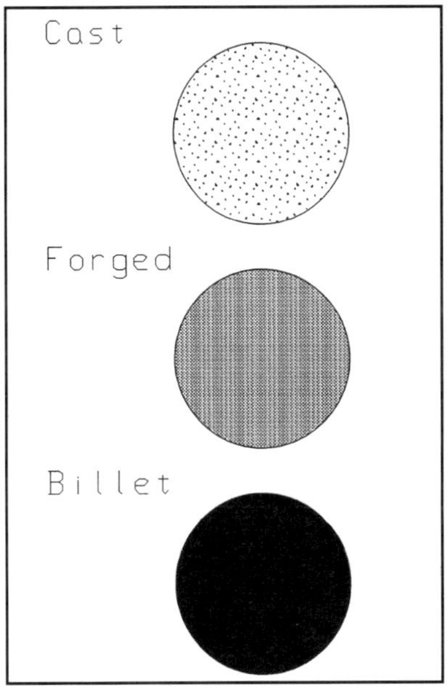

Cast cranks have long been the standard of the industry. Forged cranks have a denser structure and are therefore stronger. Billet cranks are the most expensive and are recommended for engines whose maximum rpm will approach and exceed 10,000rpm.

structure than does the cast iron crank. Additionally, the "grain" of the cast iron crank is a lot like that of a sponge, virtually nonexistent. A forged steel crank has a grain more like that of high-quality oak, very organized. The denser and more organized structure is largely a function of the manufacturing process. Forging places the crankshaft-to-be under high pressures and temperatures. Forging is the process that we see the town blacksmith performing in all those old westerns. The pressure and heat aligns the molecular structure in a strong and organized pattern.

Solid steel billet—10,000rpm+

This is the ultimate crankshaft. A billet crankshaft is literally carved out of a piece of solid billet steel. This allows for maximum strength. A second advantage is that when installed with custom connecting rods, rods that have been designed specifically for use with that crankshaft design, the rod journals can be made with a smaller diameter. Smaller journal diameters with lighter rods and pistons can mean smaller crankshaft counterweights. Smaller crankshaft counterweights can mean less rotating mass in the engine. Less rotating mass can mean more power from the engine.

As you might imagine, billet and forged cranks can be very expensive and are not available for a 1963 Hillman or many other applications. Take heart, however, a standard cast iron crank has done a superb and lasting job for street performance engines for decades. In fact, the strengthening processes available for cast iron crankshafts can improve the strength of the cast iron crank so much that your street performance dollar will always be better spent on heads, pistons, rods, and balancing.

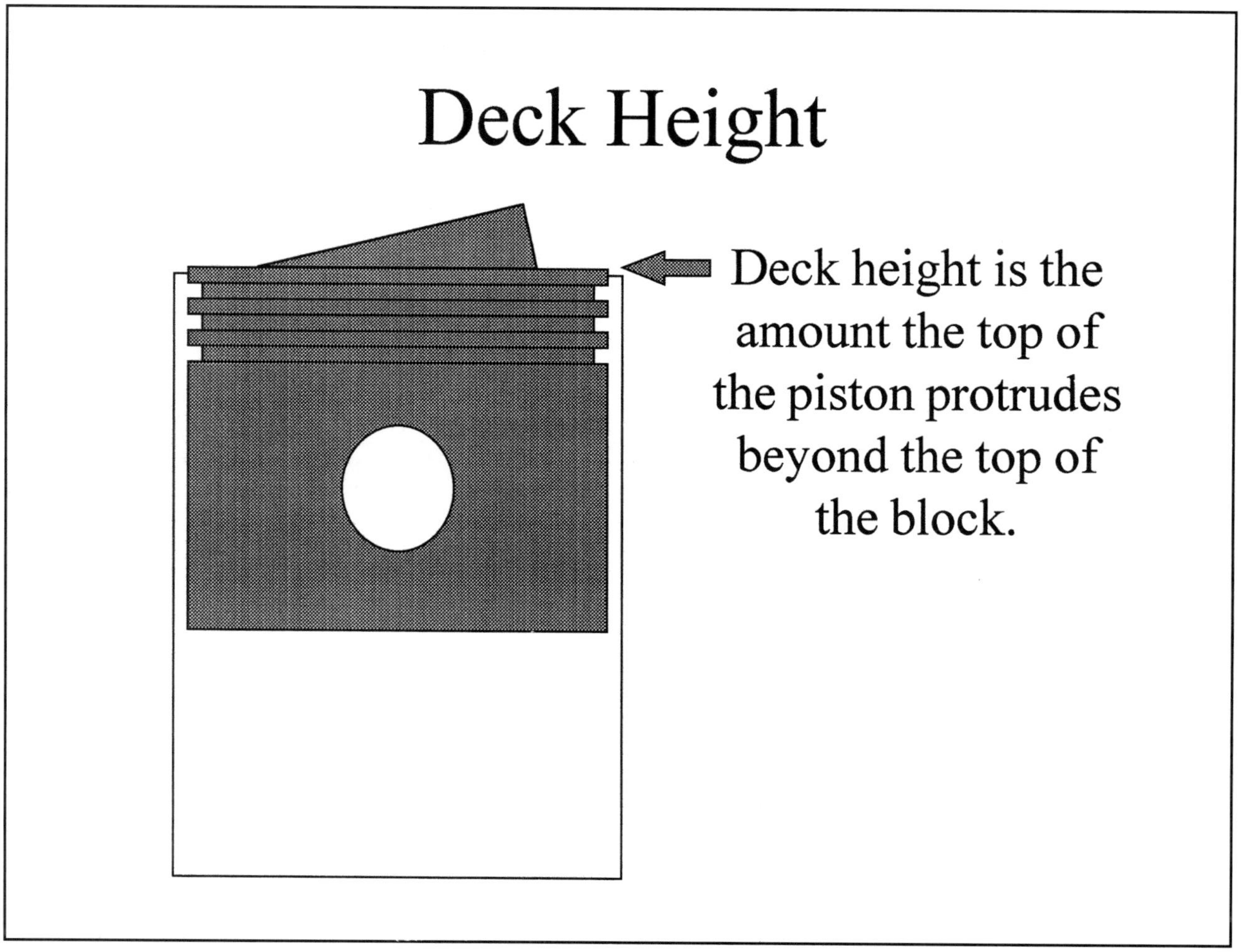

Deck height is the amount of positive or negative protrusion of the piston from the deck of the block. A negative deck height decreases the compression ratio. A positive deck height increases not only compression ratio but also the likelihood of the top of the piston contacting the bottom of the head. This not only decreases power but can cause a catastrophic failure of the reciprocating components of the engine.

Some important crankshaft facts

Deck height

Deck height is the distance of the block deck surface above the top of the piston. Actually, the preceding statement is more of a description than a definition. Deck height is the distance from the centerline of the main bearing bore to the top of the block's mating surface with the head. Deck height is an important consideration when having the block line bored. If the deck height is inadequate, damage can occur to the pistons and severely affect compression ratios. Machining operations that affect deck height are line boring and decking the block.

Line boring

This operation corrects for misalignment of the crankshaft journals in the block. Extreme heating and cooling of the engine

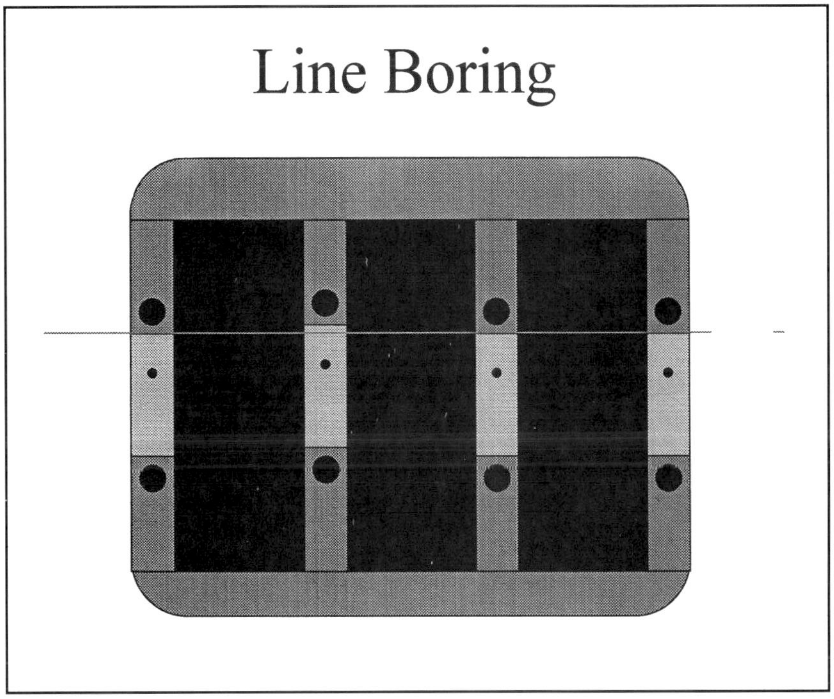

An engine block that has been overheated or subjected to a catastrophic failure may suffer from the crankshaft journals being out of alignment. A machinist will machine the ends of the bearing caps and drill a new bearing bore. This operation can affect the deck height.

Notice the casting lines at various points on this crankshaft. These are places where stress cracks and fractures can begin. Use a high-speed grinder and deburring tool to remove these lines. The crankshaft will be stronger.

Notice the slight chamfer on the oil hole of this crankshaft. This chamfering allows for better oil flow to the bearing surface. This was a standard feature on many older crankshafts. Inspect yours carefully; if they are not chamfered, use a medium-grit grinding stone to create a slight chamfer.

This crankshaft is a stock crankshaft from a late 1960s 302 Ford. The counterweights on this crankshaft are large and add considerable rotational mass. Additionally, they are not very aerodynamic. If your engine is intended for performance use, these weights should be lightened as the crankshaft is balanced. Lightening them by putting a knife-like edge on the weight not only reduces the rotational mass but increases the aerodynamics of the counterweight.

block can cause warping. Even a slight misalignment of 0.0015in (less than 1/640in) can cause a power loss and premature failure of the main bearings and possibly break the crankshaft. The block should be line bored anytime the engine is being rebuilt because of a broken crankshaft.

Line boring equipment is large and expensive. Some machine shops will opt to sublet this operation. The block is placed in a boring machine and small amounts of metal are removed. As the metal is removed, the main journals are aligned. If too much metal is removed during the boring process, the crankshaft is moved too far up in the block. This alters the compression ratio, changes the deck height of the pistons, and slackens the timing chain. If the engine has a timing chain tensioner to compensate for the slack chain, cam timing will still be retarded. If the cam timing is controlled through meshing gears, special gears will have to be purchased as the tooth penetration of stock gears will be too deep. Excessive penetration can cause premature wear. Seek the advice of your machinist on whether or not to replace the block.

If the block is badly damaged, line boring might not be an option. The block will simply need to be replaced. If the block is rare or very expensive, like those of large diesel engines, then the damaged journal boss can be sleeved and the block line bored.

Reducing the radius of the crankshaft counterweight reduces the rotational inertia of the crankshaft. This increases the ability of the engine to accelerate. Remember that as your machinist reduces the counterweight radius, he or she must also maintain balance. This is a slow and tedious procedure.

Machining and strengthening

Deburring

Like the block, a stock crankshaft will have flashing that needs to be removed. Removal of this flashing is for more than purely cosmetic purposes. The ridges formed by the flashing provides a starting point for cracks and stress fractures.

Chamfering the oil holes

Use a small, medium-grit grinding stone mounted in a drill motor to chamfer the opening of the crankshaft oil holes. Chamfering permits smoother flow of oil to the journal. The oil does not have to turn a sharp 90 degrees, and the chamfer provides a reservoir. The result is better lubrication to the rod and main journals.

Counterweights

There are several degrees to which work on the counterweights can be taken. The crankcase in which the crankshaft rotates is not a perfect environment in which to spin the crankshaft. Fine particles or a mist of oil impede the movement of the crankshaft as the oil droplets contact the crankshaft. The simplest solution, though only a first step, is to install a windage tray above the oil level in the oil pan. These are available commercially. A second step is to polish the rough surfaces on the counterweights left behind by the casting process.

If the goal is a fire-breathing quarter-mile gobbling racing engine, three additional procedures may be advisable. A popular procedure for such engines is to knife-edge the counterweights. This involves removing metal to form a 45 degree leading edge to the counterweight. In much the same way the sloped nose of a Chevy Lumina APV has better aerodynamics than a 1962 VW microbus, a knife-edged counterweight slips through the air and oil mist of the crankcase much easier. This, of course, removes weight from the counterweights and should be done only if the crank/rod/piston assemblies are to be balanced.

Counterweight radius

One of the most effective operations that can be done to the counterweights to improve performance is to reduce their radius. Removing the metal to reduce the radius lightens the crank, making the engine more responsive to speed change requests from the driver, and reduces the wind resistance that results from the counterweight swinging the air and oil mist in the block. In the ideal crankshaft, the counterweights would be near the centerline of the crankshaft. The further the weight is from the centerline, the greater the amount of energy required to change the speed of the engine. Reducing the radius not only decreases the mass but also moves the center of gravity for the mass closer to the centerline of the crankshaft.

Counterweight windage

Anybody who remembers soft drinks in bottles has probably blown across the mouth of such a bottle. The sound that is created is evidence of wind resistance. When the factory and your machinist balanced the crankshaft, they did so by drilling holes in the counterweights. These holes increase the wind resistance of the counterweight. During the final

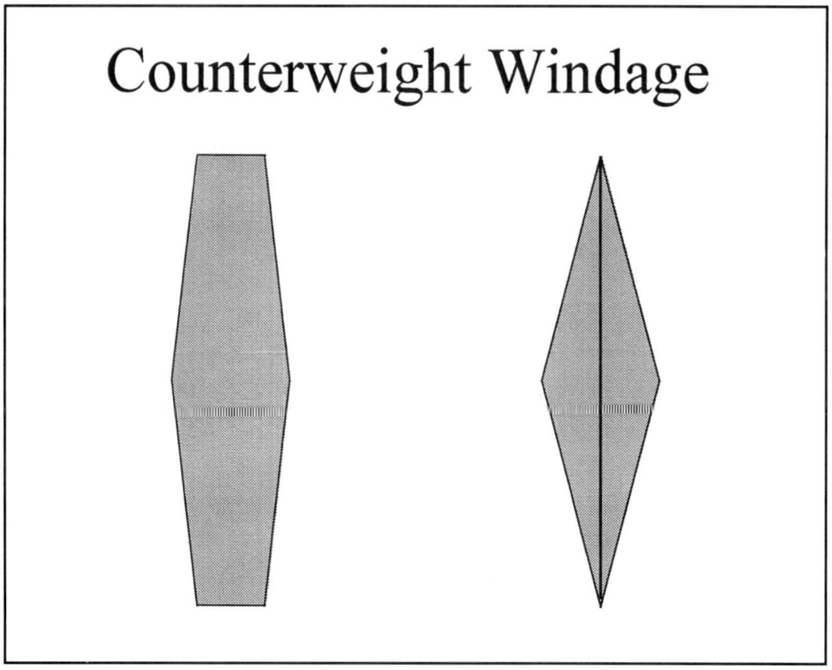

The counterweights of the crankshaft are usually blunt-edged chunks of metal. These chunks of metal slap the air to create a great deal of resistance and turbulence. If knife edges are ground on the counterweight, both turbulence and wind resistance are reduced. Remember that balance must be maintained as this operation is performed. The end result should resemble the pendulum from the old movie with Vincent Price. The one about the pit.

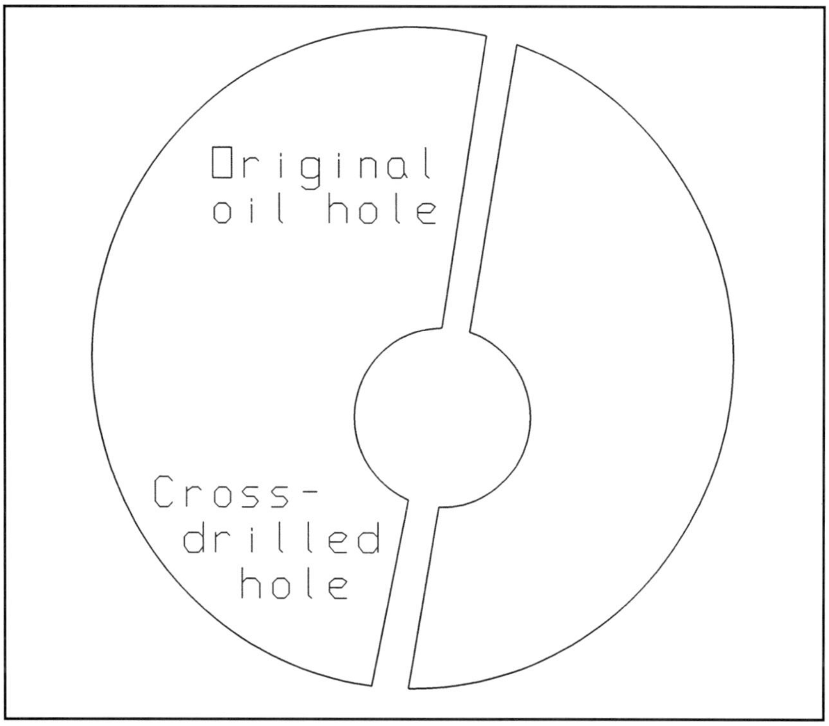

Most connecting rod journals have only one oil hole for lubricating the rod bearing. Cross-drilling the crank adds a second point of lubrication. Remember that a cross-drilled crank will require a higher volume oil pump.

stages of crank/rod/piston assembly, these holes should be closed by welding special freeze plugs over them. After these plugs are welded in place, they are machined smooth. This process is called smooth balancing.

All of the counterweight modifications you want to make should be made at the same time. Sit down with your machinist and calculate how much each procedure will cost and how many meals your spouse and children can do without before contracting what will be done.

Cross-drilling

In the typical crankshaft, each rod journal has only one hole to supply oil to the bearing. This means that oil is only delivered to one of the bearing shells at a time. Each bearing shell will only receive oil once per crankshaft revolution. Improved lubrication, and therefore bearing life, results from cross-drilling the rod journals. In this procedure, each rod journal oil supply hole is drilled straight through the journal.

Grinding

When the crankshaft has been damaged due to spun bearings or extreme wear, the crank will need to be machined. Be sure to confirm the availability of undersized bearings. Common sizes for undersized bearings are: 0.010, 0.020, and 0.030in. If working on an import application, it is especially important to confirm, as the bearings do not always run in the standard sizes.

When the crankshaft is machined to 0.020in, the machinist will mount the crank in a lathe designed exclusively for that purpose. The machinist will then set the lathe to remove 0.010in of metal from each journal of the crankshaft. Since the same amount of metal is removed all the way around, the crank will end up being 0.020in undersize. After machining the crankshaft, the machinist will then use an attachment for the lathe to polish it.

Indexing

So far the procedure is the same as during a standard overhaul. For an additional fee, the machinist can index the crank journals. As an example, on a V-8 engine this would ensure that the rod journal throws are exactly 90 degrees apart. This is accomplished by offset grinding the rod journals. Indexing ensures that the camshaft and ignition system will be exactly phased to all the pistons, not just number one.

When the crankshaft was manufactured, effort was made to ensure that each of the rod journal centerlines are the correct number of degrees apart. However, often they might be a fraction of a degree off from one another. The indexing procedure, when performed by a skilled machinist, ensures that the rod journal centerlines are exact.

Strengthening the crankshaft

Tuftriding

Tuftriding is a hardening process that involves submerging the crankshaft in a molten salt. The crankshaft is thereby heat-treated and hardened.

Shot peening and stress relief

The most common method of strengthening the crankshaft is through shot peening. After masking the journals to protect them, high-velocity shot is targeted at the other areas of the crank. Particular attention is paid to the fillet or radius areas of the main journals. The impact of the shot relieves stress in the metal that could cause microscopic cracks that could later become a broken crankshaft.

Other crankshaft operations

Straightening

If the crankshaft needs to be straightened, get a new one. However, if a lot of time, energy, or money has been put into a crankshaft, it can be straightened by a skilled machinist. This

This is the machine required to regrind a crankshaft. Do not expect this in your Christmas stocking or as an anniversary present from the wife.

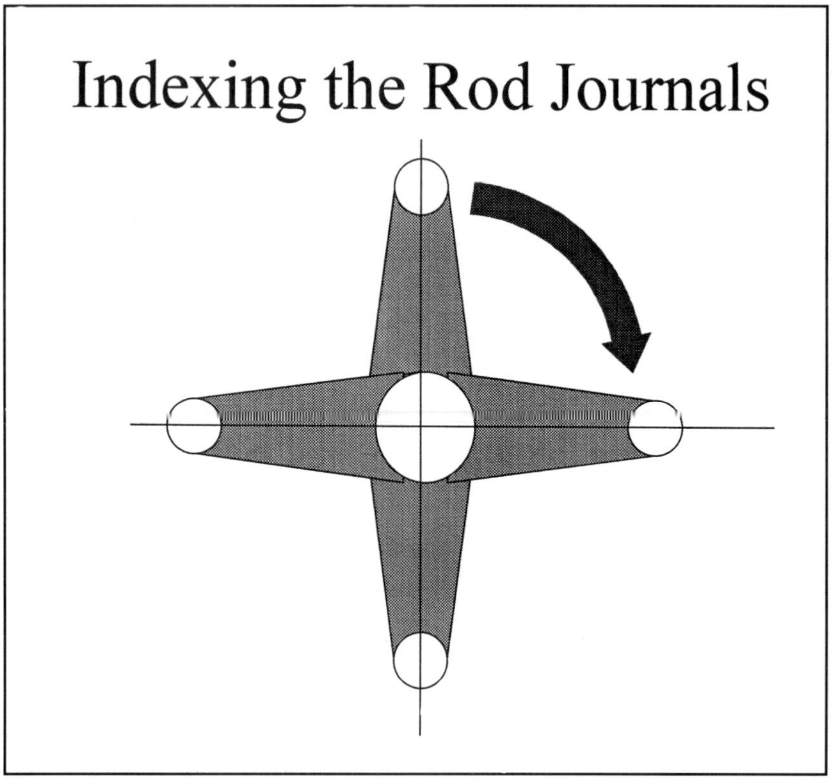

Indexing the crankshaft is a procedure that ensures the center point of each of the rod journals is exactly 90 degrees apart. When the crankshaft was manufactured, effort was made to ensure that each of the rod journal centerlines are the correct number of degrees apart. However, often they might be a fraction of a degree off from one another. The indexing procedure, when performed by a skilled machinist, ensures that the rod journal centerlines are exact.

The most common cause of a crankshaft needing to be straightened is a catastrophic failure of engine components. However, improper long-term storage can distort the crankshaft. This is the proper method for long-term storage.

is not a job for most community automotive machine shops, but they can find a place to have it done.

Polishing

Industry specifications dictate that the crank journals should be polished to a smoothness of 5 to 10 micro-inches (0.000005 to 0.000010in). Visual inspection of the crank journals to determine their need to be polished is wholly inadequate. The human eye has an ability to see objects as small as 10 microns. Ten microns is 394 micro-inches (0.000394in). The smallest object you can see therefore is forty times larger than the industry standard for smoothness.

Several methods are available for checking the smoothness of the crankshaft. The "high-tech" method is to use a profilometer. This is an electromechanical device with an accuracy 62 times more accurate than your fingernail. The bottom line is have the crankshaft polished with each engine rebuild. Polishing should also be done after the crankshaft has been machined.

Balancing the crankshaft

This is a highly specialized skill requiring highly specialized and very expensive equipment. In spite of this, the required equipment can be found in most major metropolitan areas. As I am writing this book, however, I am living in Fairbanks, Alaska, the sun fun capital of the great northwest. Challenged by the thought of writing this particular section and challenged by the curiosity of where I would get a crankshaft balanced, I made a few calls. The results of these calls was there are three places in town that can do the work on a 350ci Chevrolet crankshaft for less than $100. A forged crank would be a little more expensive.

Main bearing journals should be mic'ed for out of round and taper before a used crank is installed.

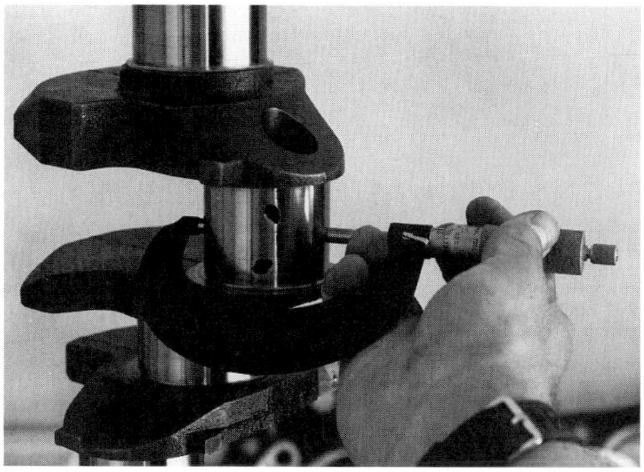

Connecting rod journals should be mic'ed for out of round and taper before a used crank is installed.

Before installing the crank bearings, the journal boss of the block should be inspected for dirt and grit. Lay the bearing in the boss and slide it into position with your thumbs. Be sure that the tab on the bearing aligns with the notch in the journal boss.

If there are any balancing mechanisms that are external of the engine, harmonic balancers, flywheels, and so on, these components must be supplied to the machinist balancing the crankshaft. The procedure begins with installing bobweights on the crankshaft. The crank is then rotated at a high rpm. At this point, some high-tech crankshaft balancing machines will give you an exact readout of the points where the removal of metal is required and how much metal to remove. Other machines operate through the slow and tedious method of trial and error.

Note that one of the bearings will have a thrust bearing surface. This is what will determine the crankshaft end-play.

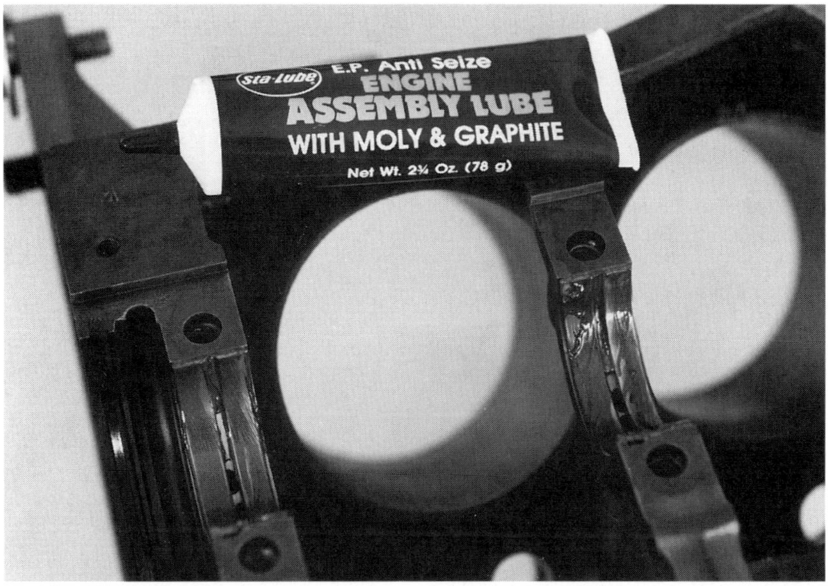

Lubricate the bearings with a high-quality engine assembly lube. This will protect the new engine bearing when the crankshaft is being rotated or cranked prior to establishing oil pressure.

Roller cranks

There are a few specialized applications that replace the traditional oil film separation sliding bearing with needle bearings. I first heard of this with respect to the Volkswagen performance builders of the 1960s. Those of you who first became aware of popular music when John Travolta was wearing a white leisure suit may consider "Volkswagen performance" to be an oxymoron. Yet, some of the most sophisticated lower end performance work ever done on a stock engine was done on the 1200 and 1300cc VW engines.

The advantages of roller bearings are obvious. The bearing surfaces themselves, each one of the needles, is much tougher and more durable than the obligatory soft surface of a hydrodynamic sliding bearing. The unlubricated rolling resistance is much lower with roller bearings. The dynamic frictional forces of the crankshaft at speed are greatly reduced with the crankshaft at speed. This reduces the vibration factors discussed earlier in this chapter, which in turn, reduces crankshaft loading. The bottom line: increased maximum engine speed with reduced risk of damage.

Stroker cranks

Many years ago I was a junior high school student at an American high school in London, England. This was during a time when most of the United States culture was worshipping the "surf, sun, and cars" culture of California. In the United Kingdom, we found it a rare treat to get to hear an occasional Beach Boys song over one of the pirate radio stations. Such an occurrence would be virtually impossible on the BBC. It was at this time, while perusing the magazine racks at the Stars and Stripes bookstore at South Ruislip Air Force Base, that I found my first *CARtoons* magazine. In this first issue I had ever seen,

Lay the crankshaft into the bearings gently. You have probably spent a great deal of time and effort ensuring precision up to this point; do not ruin it by hasty assembly practices.

one of the characters mentioned that his engine had been bored and stroked. I, of course, did not find out what this meant for many years.

The typical modern V-8 engine has a typical stroke of 3 to 4in. While this stroke provides a great deal of leverage, and therefore torque, it limits the maximum rpm. The longer the stroke, the greater the piston speed at the midpoint of its travel. The greater the piston speed, the greater the possibility of the piston self-destructing.

A stroker crank allows the speed of the engine to be increased to a point where maximum horsepower can be obtained from a given cubic centimeters displacement. To exaggerate this a bit, let us say that we are planning to campaign a

Torque the main bearing caps to their proper specifications. Do not guess; for this specification, use the factory service manual where available or third-party sources.

Crankshaft end-play is a critical measurement. After all the main bearings are torqued, check the crankshaft end-play. Install a dial indicator on the block to detect movement in the crankshaft. Using a pry bar, gently force the crank as far back in the block as possible. Adjust the dial indicator to zero and pry the crank as far forward as possible. The dial indicator will show crankshaft end-play. Allowable end-play differs from one application to another. An alternative to the dial indicator is to use a feeler gauge. A general rule of thumb is 0.003 to 0.008in. If the end-play is incorrect, the crankshaft thrust surfaces will need to be repaired.

car in a class that limits cubic centimeters to 1,000. We could use an engine that has four cylinders that has a stroke of about 2.7in and a bore of about 2.7in. Or, as an alternative, we could use a V-8 engine with a bore of 4in and a stroke of about 1.55in. When this V-8 engine had 3in of stroke, the estimated maximum rpm would be about 7000rpm. With a stroke of 1.55in, the V-8 engine would have a maximum engine speed of over 13,000rpm. Even the little four-cylinder engine would only have a maximum piston speed of about 7700rpm. In both cases, the allowable piston speed has been considered to be 3500rpm.

This example of stroking the engine may have increased the

Below
When the main bearings are properly torqued, the crankshaft should rotate easily by hand.

maximum engine speed; however, this was at the cost of the compression ratio. If the original 5.0ltr version of the engine had a compression ratio of 10.5:1, then when the engine was stroked the compression ratio dropped to 5.9:1. Of course, the compression ratio could be brought back to its original 10.5:1 if the combustion chamber volume was decreased from 65cc to 33.5cc. Of course, to decrease combustion chamber size by this amount, it is virtually necessary to do so by using domed pistons. Domed pistons increase the likelihood that there will be an interference in the flow of the flame front through the combustion chamber. This interruption, as previously covered, will decrease the efficiency of the combustion and reduce power.

Of course, this example goes beyond the ridiculous. It does, however, illustrate the concerns you must analyze when you consider stroking an engine. Typically, stroking the engine is a method of making a large engine smaller so that it can be used in a certain competition class or to increase rpm capability of the engine without increasing the piston speed.

Stroker Cranks

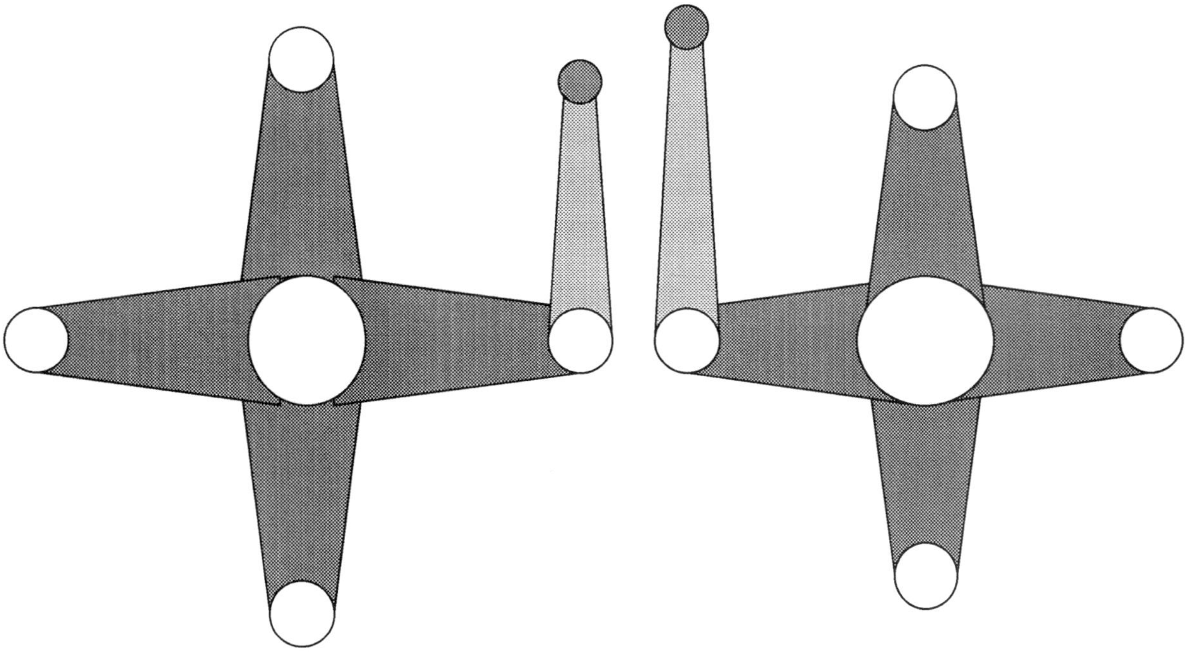

The illustration on the left represents a standard crankshaft. The illustration on the right represents a stroker crank. Stroker cranks allow higher engine speed without increasing the piston speed. This higher maximum rpm is achieved at the sacrifice of compression ratio. Notice that the connecting rod used with the stroker crank is longer. Because the rod journal at top dead center is further from the combustion chamber of the head, this difference in distance must be made up for by a longer connecting rod.

Chapter 6
Bearings, Rods and Mains

Types of bearings
Hydrodynamic sliding bearings

Hydrodynamic sliding bearings is a fancy name for the oil film bearing found in most modern gasoline and diesel engines. These bearings depend on two smooth surfaces sliding past one another separated by a film of lubricating fluid. Because there is no direct contact between the two smooth metal surfaces, wear is kept at a minimum. To give you an idea of what a good bearing the hydrodynamic sliding bearing can be, a surface-to-surface, dry friction bearing can have a coefficient of friction ranging from about 0.1 to greater than 1. A roller bearing will have a coefficient of friction of about 0.001. A fluid friction bearing, the type used in internal-combustion engines, will have a coefficient of friction of about 0.01. While this is higher than the roller bearing, it is still extremely low.

Multilayer bearings

Connecting rod bearings and crankshaft main bearings are subjected to extremely high dynamic loads. The severity of these loads is considerable. Most engine bearings, therefore, consist of layers of different materials, each material adding a different benefit and strength. Engine rod and main bearings have many criteria they must meet to perform their job well.

Conformability

Were we living on the perfect world, conformability would not be an issue. Comformability is the ability of the connecting rod or main bearing to adapt to minor misalignment without reducing the service life of the bearing. Although a harder, stiffer bearing might wear longer, it would require perfect mating surfaces.

Wetability

I am sure that you have either waxed a car or seen a car that has been recently waxed. You have probably noticed that when the freshly waxed surface is sprayed with water, it stands in beads on the surface. This is because the waxed surface has poor wetability. When I was a kid, on the other hand, my dad had a 1953 Chevrolet pickup. This truck was powder blue. Actually it was dark blue, but so badly oxidized that when you rubbed your hand across the surface it would be covered with blue powder. When it rained or

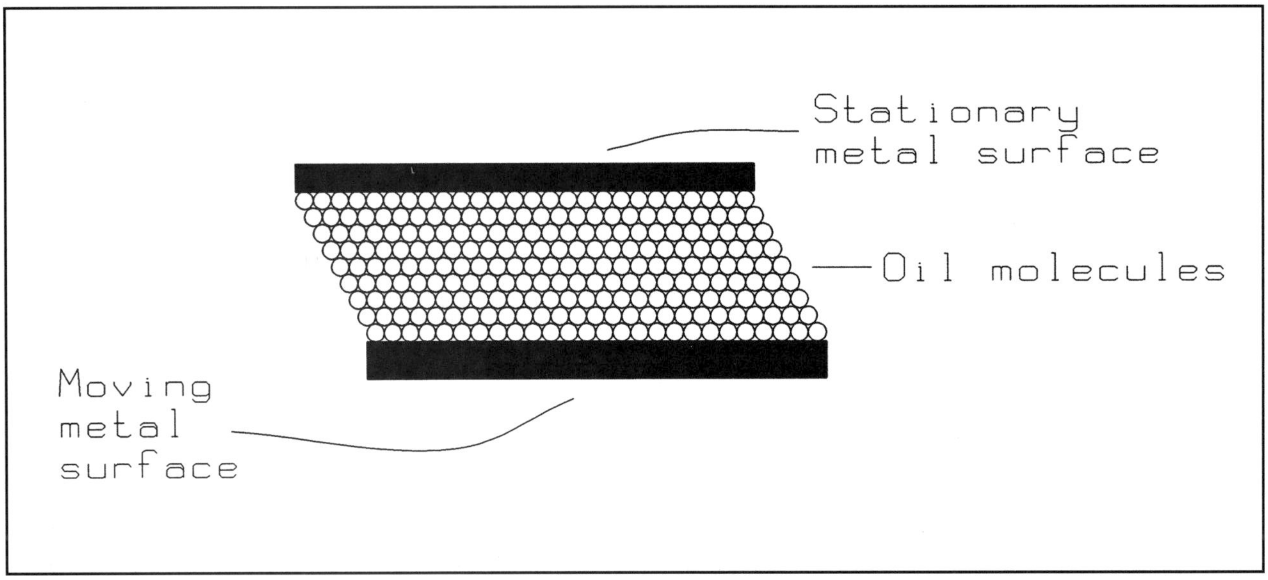

When the engine is running, the bearing and rod or main journal should never contact one another. A thin layer of oil is supposed to provide separation. This is why good lubrication, oil volume, and oil pressure are so important. In reality, it is not the bearing but the oil that is the real low-friction contact between the bearing and the crankshaft. This type of bearing is known as a hydrodynamic bearing.

when the surface of the truck was sprayed with water, it would sheet across the surface without beading. This is what a good hydrodynamic sliding bearing should do.

Embedability

Even the cleanest engine oil carries small fragments of metal, dirt, dust, and grit to the bearing surfaces. If these particles are larger than the oil clearance between the bearing surfaces, then an "interference fit" occurs. Something has to give. The most desirable thing to give is the particle. However, if it is harder than either the bearing surface or the crankshaft journal, the latter will have to absorb it. Since crankshafts are more expensive than bearings, it is desirable that the bearing surface be malleable enough to absorb the particle.

Wear resistance

This is probably an obvious requirement of a bearing. Today's engines are expected to last well over 100,000 miles, maybe even 200,000 miles before major work is done. In this time, each main bearing has had the crankshaft spin inside of it over 400 million times. One would hope that in this time the metal surface of the crankshaft and the surface of the bearing have made only very infrequent physical contact with one another.

Seizure resistance

This factor has to do with the bearing's resistance to welding itself to the rotating shaft. The bearing must resist this temptation even under the intense persuasion of extreme compressive loads and high sliding velocity.

Run-in performance

The hardest job most of the engine components will ever have is to service the initial post-rebuild start-up. During this time lubrication is poor,

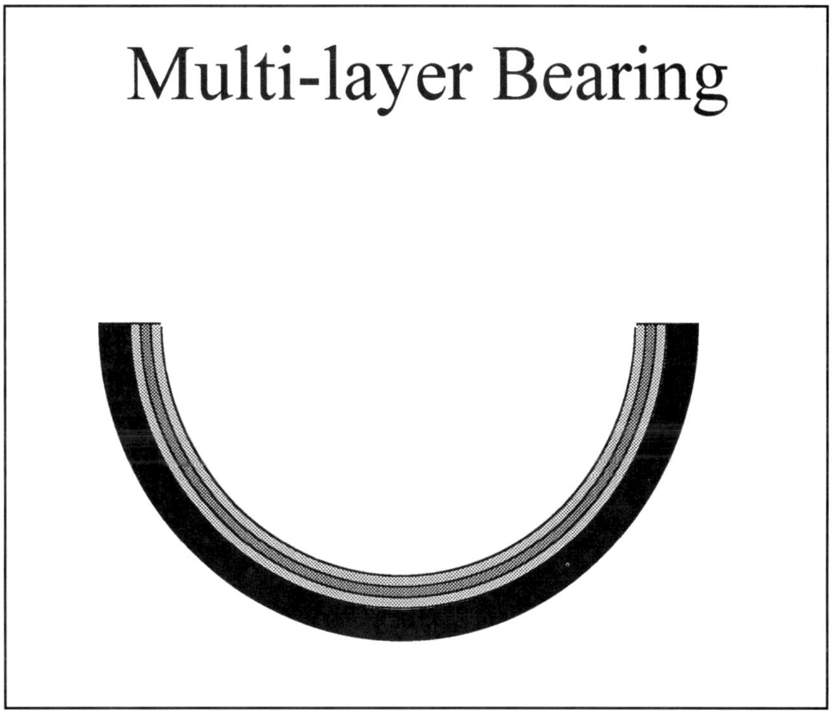

Most modern engine bearings are multilayer bearings. There are a wide range of needs and requirements that engine bearings need to meet. No single material can provide maximum capability in each of these areas. Bearing manufacturers will use several materials, taking advantage of the best characteristics of each material.

Bearing Design Considerations

- Conformability
- Wettability
- Embeddability
- Wear Resistance
- Siezure Resistance
- Run-in Performance
- Mechanical Loadability
- Fatigue Strength

This bearing surface is an example of one of the required characteristics of a bearing. In the center of the bearing, there is an area that appears to be a fingerprint. In reality, this is the remains of dirt and grit that were not cleaned off prior to assembly. The material has been imbedded into the bearing and is therefore isolated from direct contact with the crank journal.

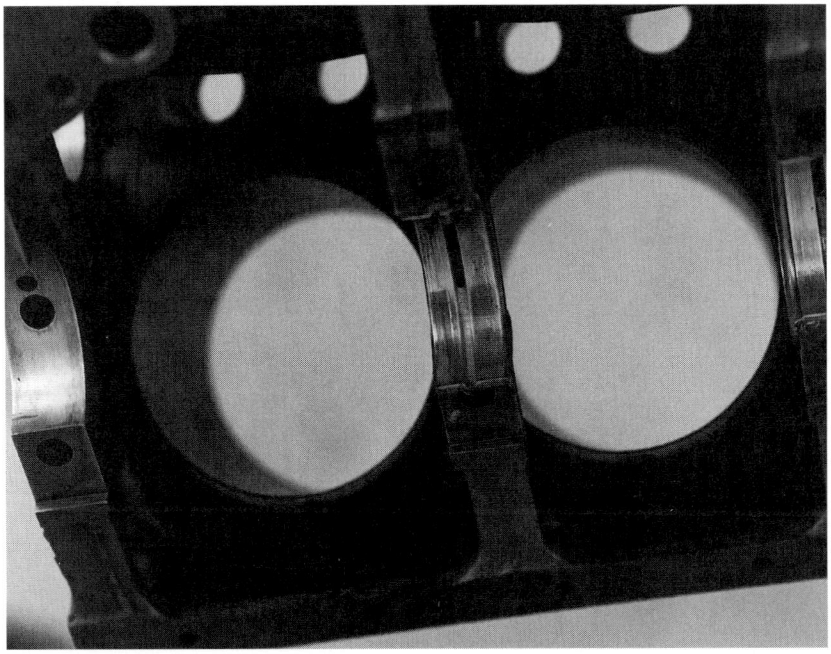

conformability factors are at their maximum point of demand, and any of the bearing surfaces that have not been perfectly cleaned of grit will have to allow that remaining grit to imbed. Oddly, the most rapid damage to bearings occurs when the engine is first started and even prior to the first start as the crankshaft is being rotated by hand.

Left
These bearings were installed in a small block Ford engine over twenty years before this picture was taken. In spite of their antiquity, they show little evidence of wear. Good maintenance, regular oil changes, limited abuse, and good bearing wearability have all served to make this possible.

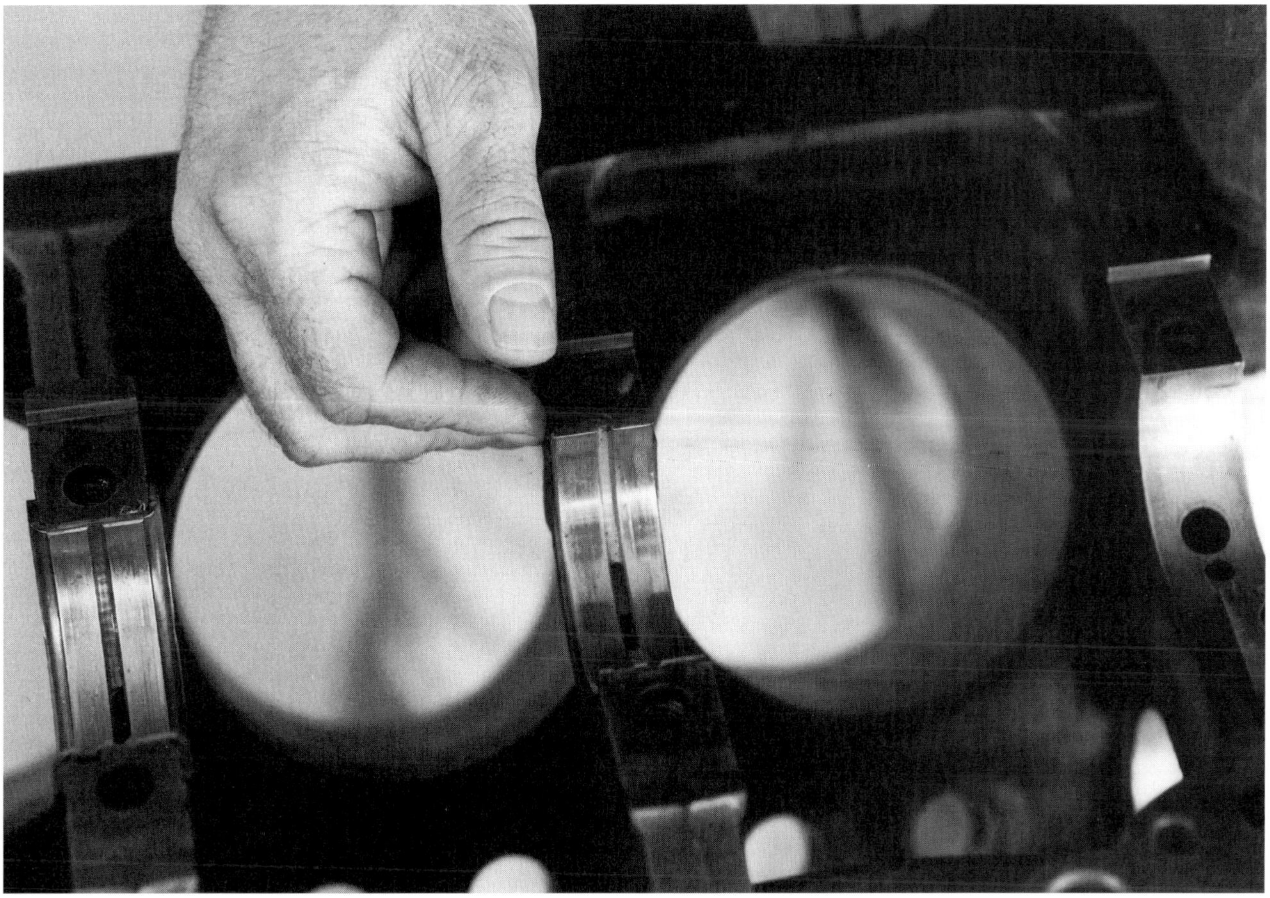

These bearings are relatively new, less than 500 miles. Notice that they appear to be worn in a similar manner to the twenty-plus-year-old bearing previously discussed. This is because there is an outer coating on the bearing specifically designed to be worn away in the first few hours of operation. This bearing material excels in its run-in performance.

Mechanical loadability

If the bearing materials are too soft, they will have good embedability but could be "mooshed" out. The inability to accept mechanical loading will cause the oil clearance to be hammered larger. Even though the oil film between the crankshaft and bearing forms a good cushion, some loading will be transferred throughout the oil film to the bearing. The bearing must be hard enough to withstand these loads.

Fatigue strength

Remember that the rod and crankshaft bearings are subjected to day after day of loading and pounding. Remember that we expect this bearing to be able

What is truly amazing is that the relatively soft bearing material is capable of withstanding the same load conditions that the cast iron crankcase was designed to handle.

The cam bearings are often ignored by the lesser professional. Rumors that the cam bearings never wear dominate the industry. No overhaul should occur without replacement of these bearings.

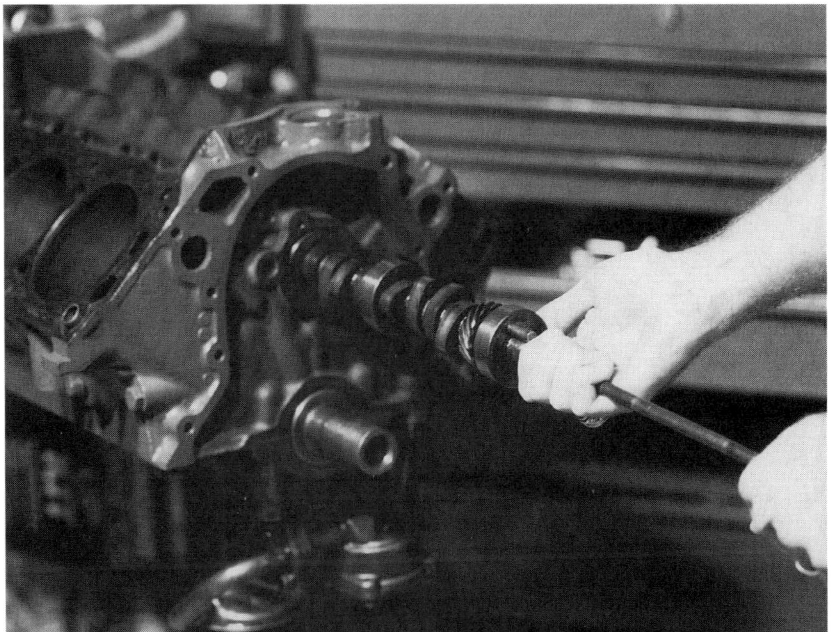

Use extreme care when installing the camshaft. Damaging the cam bearing during camshaft installation is a common problem.

to withstand this pounding hundreds of millions of times in the life of the engine.

The bottom line

An engine bearing must be able to perform an almost impossible task of compromise, a task not unlike trying to get 535 politicians to forget their personal agendas and do what is best for the country. The bearing must be pliable enough to have good conformability but not so pliable as to have poor loadability. The bearing must have good loadability characteristics and yet be soft enough to have good fatigue strength. It must have good embedability yet good seizure and welding resistance. These are difficult and conflicting requirements for a bearing, and only a precise

Material	HB Hardness @ 100∞ C	Remarks
Tin-based babbit metal	10	very soft, good conformability, excellent anti-seize performance
Lead-base babbit metal	9	reinforcement necessary
Lead-base bronze	47	very soft, good anti-seize performance, low resistance to wear
Lead-tin-base bronze	67	excellent anti-seize performance, good resistance to off-axis operation, this makes excellent crankshaft bearing material
Tin-base bronze	85	hard material, best for high load conditions with low sliding velocities
Red brass	65	best for use with moderate loads, suitable for use as wrist pin bearings, bushings, and crankshaft bearings
Brass	90	Best for low temperature, low load use
Aluminum bronze	110	very hard, high wear resistance, good in interference fit applications, increased friction
Aluminum alloy	100	piston alloy for low sliding velocities
Rolled aluminum cladding	30	high loadability, good anti-friction properties

combination of materials can meet all the requirements.

Materials

Refer to the chart at the top of this page for information on materials.

Installing bearings

Rod bearing and rod journal condition

Like the main bearings and journals, the rod bearings and journals can cause a loss of oil pressure if the clearance between rod bearings and the rod journals of the crankshaft increases dramatically. The increase in the clearance allows oil to easily escape from the space between the bearing and journal. The escaping oil causes the oil pressure to decrease. The only repair for this problem is machining or replacing the crankshaft and replacing the bearings.

Well, you probably thought you were done with the tedium of measurements. All the parts are back from the machine shop and you are ready to reassemble. Sorry, it is time to get those micrometers, dial indicators, and feeler gauges back out. Confirming the many measurements as the new parts are installed is part of the reassembly process.

Turn the block upside down. Clean the main bearing saddles thoroughly with a light solvent. The oil clearances of the main bearings will be less than 0.002in. Even a small piece of sand could distort the bearing and bind the crankshaft. Place the main bearing shells in the saddles. Be careful to note two things: First, install the main bearing with the thrust surface on the correct saddle; and second, if only half the bearing shells have an oil supply hole in them, be sure that the hole lines up with the oil gallery port. Take it from experience, the engine will make an annoying knocking sound if the oil cannot get through the bearing to the crank journal. With the bearing shells properly laid in the saddle, smear the assembly lube across the bearing surface. Carefully lay the crankshaft in the block.

Before installing the bearings in their respective journals, make sure that the mating surface is clean. A small piece of dirt or grit can cause improper bearing clearance when the bearing cap is torqued.

Lay a short piece of plastigauge on each of the crank journals. Clean the main bearing caps to remove grit and place the bearing shells in them. Install the main bearing caps and torque them to the proper specification. Remove the bearing caps and use the plastigauge packaging to measure its width on the journal. Typical bearing clearances for passenger car and light-duty truck engines are 0.0015 to 0.002in. If the bearing clearance is incorrect, you must have measured your crank diameter incorrectly, you have the wrong bearings, or there was a mistake in the machining process. Remove the bearing shell from the bearing cap. On the back side of the shell, near one end, there should be a small marking indicating the undersize of the bearing (e.g., .010 for ten-thousandth undersize). Be sure this marking coincides with the marking the machinist stamped on the crankshaft near one of the main journals. If they match, mic the crank; it was probably machined to a size other than that indicated.

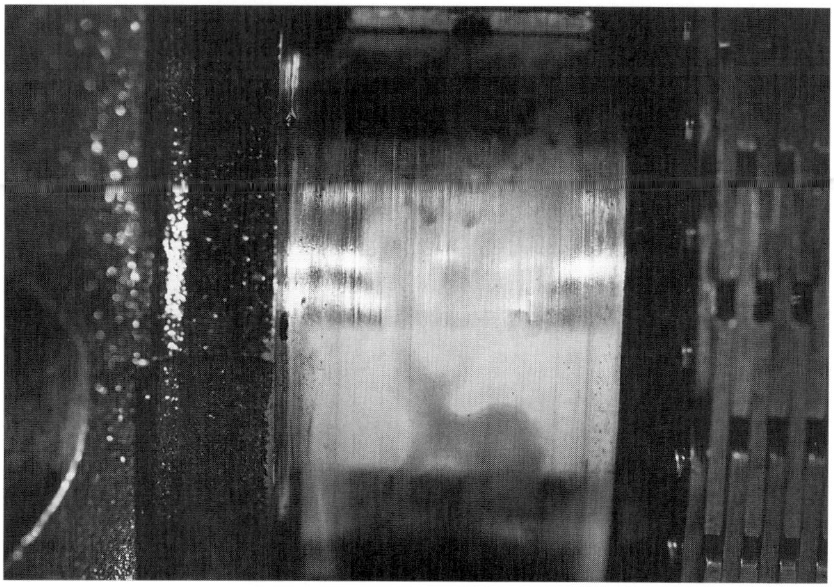

This journal was not cleaned properly before the cap was torqued. The engine was run less than 500 miles before being disassembled again. Even though the embedabillity characteristic of the bearing material absorbed part of the foreign material, damage was still done to the crankshaft. This journal will now have to be polished.

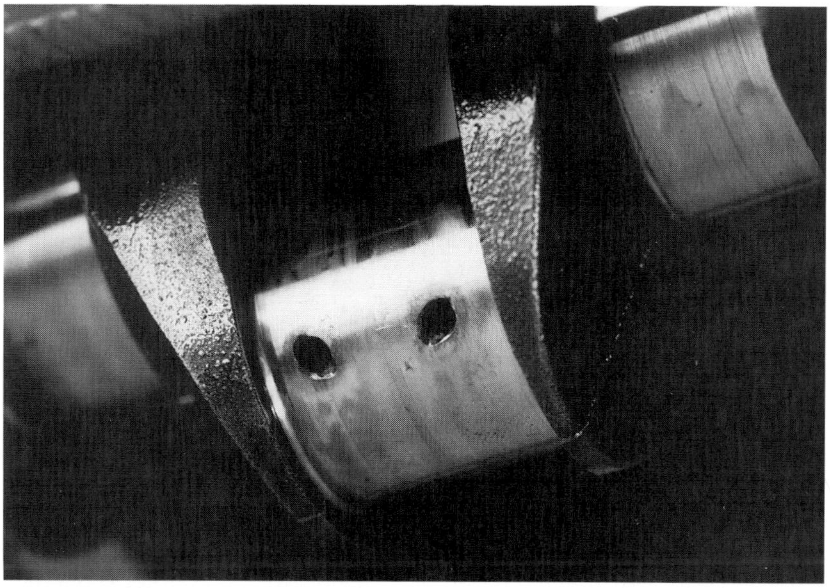

These rod journals were not cleaned properly, either.

If the plastigauge shows the oil clearance to be correct, apply assembly lube and install the main bearing caps and torque them to the proper specification. Some people like to remove the plastigauge before installing the bearing caps; it is not really necessary.

After all the main bearings are torqued, check the crankshaft end-play. Install a dial indicator on the block to detect movement in the crankshaft. Using a pry bar, gently force the crank as far back in the block as possible. Adjust the dial indicator to zero and pry the crank as far forward as possible. The dial indicator will show crankshaft end-play, although allowable end play differs from one application to another. A general rule of thumb is 0.003 to 0.008. If the end play is incorrect, the crankshaft thrust surfaces will need to be repaired.

If you do not have access to a dial indicator, an alternative, though slightly less accurate method, is to use a feeler gauge. Pry the crankshaft all the way to the rear and place the largest feeler gauge blade possible between the crank thrust surface and the main bearing thrust surface.

Chapter 7

The Crankcase, the Block

Inspection

Crack inspection

Even the most discerning individual can miss cracks that give themselves away only as the slightest discoloration. More scientific methods are available, methods that can even detect cracks hidden below the surface. This technique is called magnafluxing. The block is placed on a powerful electromagnet. Fine iron powder is sprinkled on the block. The magnetic field will align the iron powder in a distinctive pattern. A crack will interrupt this pattern. When the technician sees this interruption in the evenness of the powder, he knows that a crack is present.

Dye testing

Have you ever gone into one of those roadside diners along Route 66 in northern Arizona? Few people know that all of the cracked coffee cups in the universe are transported there by supreme forces beyond our comprehension. If you examine one of the cracks in the cup you will find there is residue from the coffee of customers who sipped a little java while waiting for the bus to take them off to World War II. This is the principle of dye crack testing. A penetrating dye is sprayed on the surface of the item being tested. After being allowed to soak in and dry, a cleaner is used to remove dye from the surface. Like the coffee stains in the crack of the cup, the dye will remain in the cracks of the component being tested. The dye method is especially effective on nonferrous metals such as aluminum and magnesium.

The first thing to look for when choosing a block is cracks. This cylinder head shows a crack in a common location for both heads and blocks. These stress-related cracks are often found running between water jacket passageways. These are common in climates like the one where I live. Here in Fairbanks we can expect several weeks every year with temperatures staying at less than -30 degrees F.

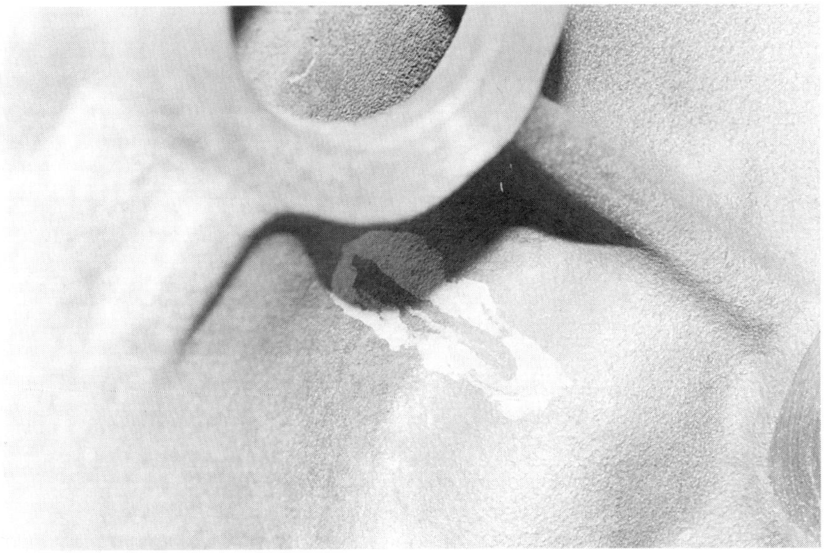

The crack in this block was missed during careful inspection by two experienced technicians. What is ironic is that one of the technicians had witnessed a water leak in this area before the engine was removed from the car. Only dye testing was effective in locating the leak. The technician now has a choice to make. Does he replace the block or have it stop drilled and welded?

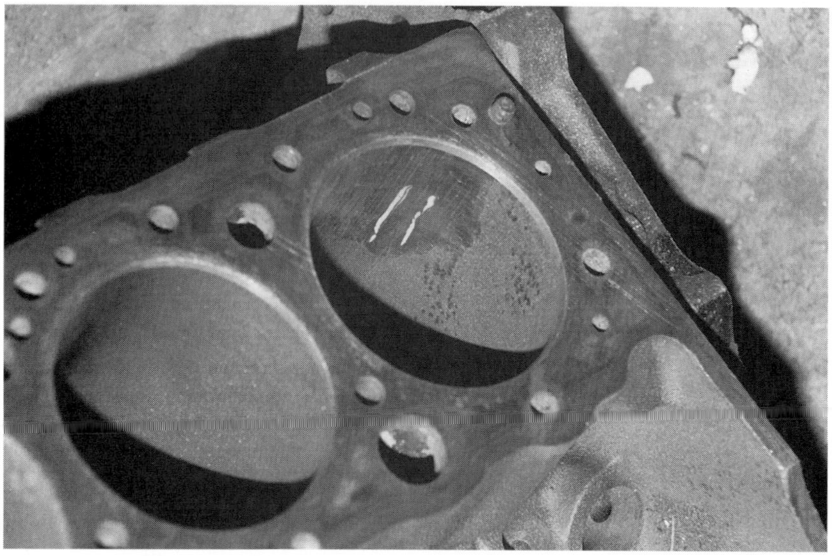

Cracked cylinder walls are common.

Look carefully at this head. There is a crack just to the left of the word "crack." It runs between the head bolt hole and the combustion chamber. Many cracks in the deck of the block will be just as hard to see.

Sonic testing

Diesel engines that use wet sleeve cylinders have long suffered from a problem called cavitation erosion. During combustion the high pressures in the cylinder cause the cylinder wall to bow outward. As the burn in the cylinder continues, the pressure begins to drop. The bowed cylinder liner then pops back to its original position. For the briefest moment, a vacuum occurs around the perimeter of the liner. The cylinder has returned to its normal position and shape, but it takes a moment longer for the coolant that surrounds the cylinder to return to contact with the cylinder. As the coolant slaps back up against the cylinder, a microscopic amount of metal is displaced from the cylinder's outer surface. This process occurs hundreds of thousands, if not millions, of times in the life of the engine. Eventually the cylinder wall can be weakened to the point of catastrophic failure.

A similar condition can occur in gasoline engines, especially performance engines. A thin cylinder wall can occur from casting problems or because the block has been bored. While boring the block does increase the displacement and the compression ratio of the engine, it also weakens the cylinder walls. In a high-performance engine, MEP can be high enough to flex the cylinder walls to the point of compression loss. Any compression loss will, of course, cause a loss of power. Even in engines built with a lower MEP, long-term cavitation erosion can still occur.

One of the most effective ways to measure the thickness of the cylinder walls is with an ultrasonic tester. Since ultrasonic testers are rather pricey, you probably will have this operation done by a machine shop. This minimum cylinder wall thickness should be greater than 0.150in for most engines. Frankly, in most cases, the weight of the engine is of little or no concern to the builder. In order to keep the compression loss to a minimum, the thickness of the cylinder walls after they are bored should be greater than 0.180in. Big block high-performance engines should have a cylinder wall thickness greater than 0.200in.

Repairing a cracked block

Cracked blocks are common in some applications, climates, and uses. They are rare in others. If the block is cracked, there are several repair methods available to you, welding and threaded plug among them.

Welding the block is a highly specialized art; most machinists recommend replacement. If the block is a rare application, consult your machinist for recommendations on who is qualified locally to do the welding.

Sealing the crack with threaded plugs works well when the crack is small and replacement or welding of the block is out of the question. In this technique, the machinist will stop drill the crack, then drill a series of large holes along the length of the crack. Special threaded plugs are then threaded into the holes. When threaded as far as they will go, they are cut off and ground flush with the block surface. Consult your machinist for advice on repairing cracked blocks.

Deburring

This is an operation that can be performed by almost anyone. Anyone who has built a model car or model airplane is familiar with the flashing that results from seepage of the plastic into the joints between the sections of the mold. When building a plastic model, this flashing must be removed to make the final product look neat.

In an engine block, there is flashing that marks the junction between the pieces of the mold. Removal of this flashing is for more than purely cosmetic purposes. The ridges formed by the flashing provides a starting point for cracks and stress fractures.

Mount a rotary file in an electric drill motor. Air drills are acceptable if the motor has a variable speed. Most single speed air drills spin so fast that the unwary hand could accidentally grind

This is a pretty engine block. The builder has effectively used a spray paint can to give the appearance of quality. If you look closely, there are casting marks, imperfections, and ridges everywhere. These casting imperfections are good starting points for cracks.

away the block and leave the flashing behind. Remove any flash evident on the interior of the crankcase or the intake valley. Other places to look include the drain holes to the crankcase. Do not get carried away with this procedure. The idea is to remove the flashing, not polish the inside of the block. While a polished block may seem like a nice idea, in reality the removal of all the little dimples reduces the surface area of the block and reduces heat transfer from the oil to the block.

There are places in the block where the rotary file will not fit. A flat-tipped punch will work well in those places. A file should be used to debur the edges of the main bearing caps. Deburring the lifter holes can be done with a wheel-cylinder abrasive ball hone. It is not advisable to use a flat stone wheel-cylinder hone for this purpose.

Continue the deburring process until all the flashing is removed.

This engine block, however, has been dressed out nicely. All casting lines and burrs have been removed. The grain of the casting has been virtually eliminated. This block is ready for the stresses of racing.

Decking

Decking the block makes sure that the mating surface with the head is flat and even. It really makes no sense, especially in a performance engine, to pay money to have the cylinder heads machined flat and then install them on a distorted cylinder block surface. This process, when done properly, also ensures the angle of the head mating surfaces is correct. For most "V" engines, this angle is either 60 or 90 degrees. The accuracy of this angle is important to allow proper fit of the intake manifold to the cylinder head.

For most performance-minded individuals, the purpose of decking the block is to provide the proper clearance between the piston at the top of the stroke and the bottom of the cylinder head. Without giving it a great deal of thought, one might feel that the piston should be flush with the top of the block when it is at the top of its stroke. This would be true if the metal of the connecting rods were not slightly elastic. The piston is an inertial mass moving at approximately 2,000ft per minute (nearly 23mph) as it travels up the cylinder. As the piston reaches the top of its travel, it must stop and reverse direction in as little as 3 milliseconds. During that time there are tremendous forces working on the connecting rod. This causes the connecting rod to stretch. When building an engine for street use, the static clearance between the top of the piston and the deck should be

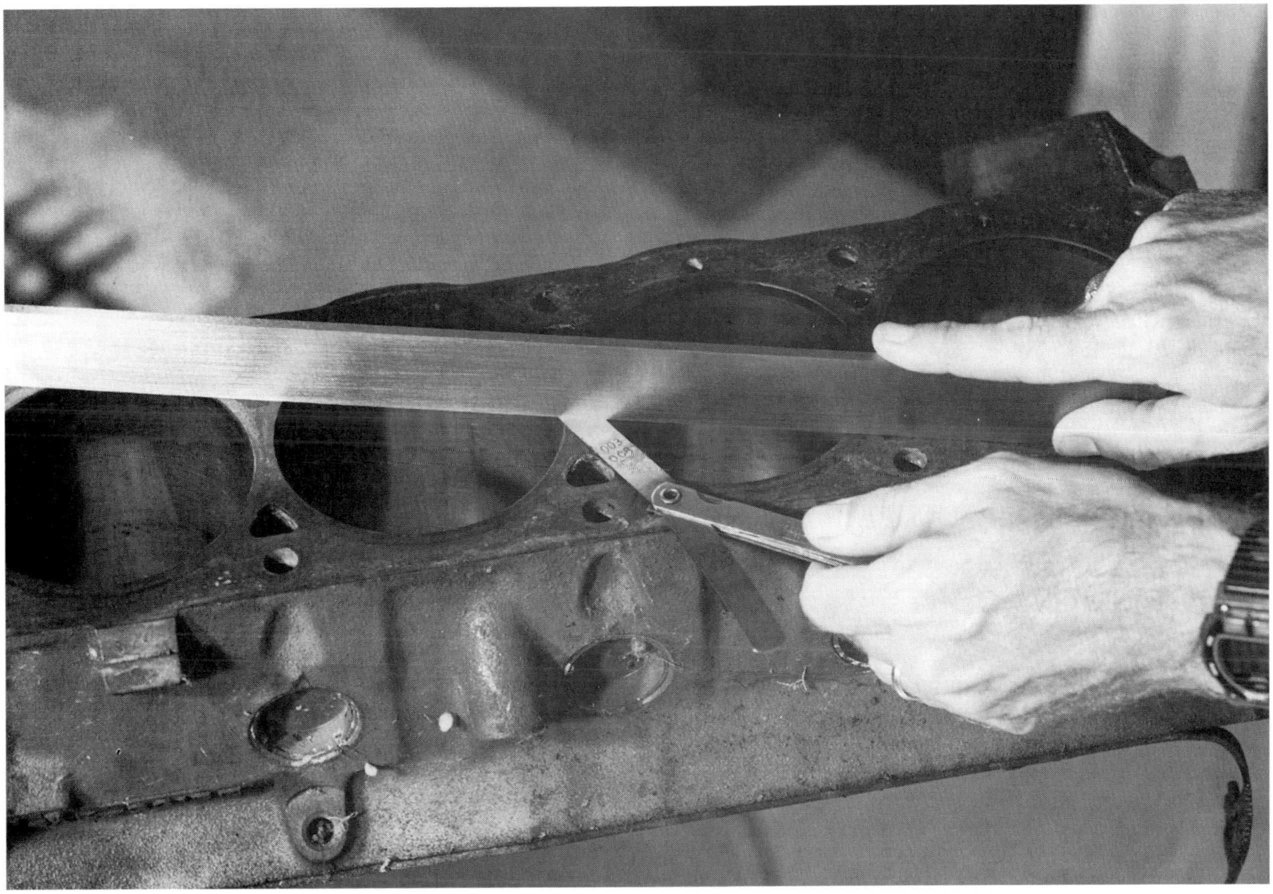

Although cylinder heads are much more prone to distortion than is the block, the deck or cylinder head mating surface can also become distorted. Use a machined straight-edge. This engine has 0.003in warpage.

0.035in. Due to potentially much higher engine speeds, this clearance should be 0.045in on racing engines. If the engine is being reassembled with aluminum connecting rods, the clearance should be 0.060in.

When deck-to-piston top or clearance is being checked, you

Right
Remember that resurfacing the deck of the block will affect the deck height. When building an engine for street use, the static clearance between the top of the piston and the deck should be 0.035in. Due to potentially much higher engine speeds, this clearance should be 0.045in on racing engines. If the engine is being reassembled with aluminum connecting rods, the clearance should be 0.060in. This engine already has a negative deck height of 0.050in. This block should just be replaced.

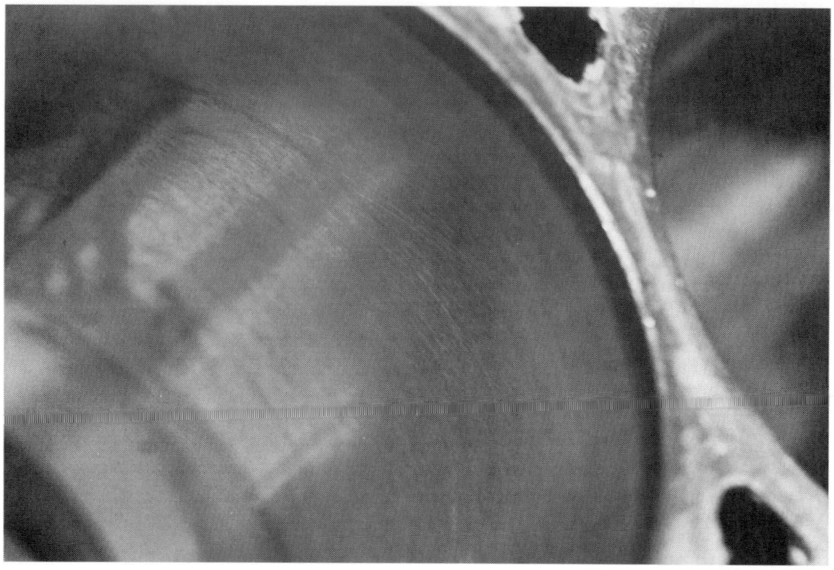

Honing breaks the glaze and gives the rings an abrasive surface to help seat them. When blueprinting an engine, honing also ensures that the cylinder's walls are as round and straight as possible.

should consider the thickness of the head gasket. If you are building a stock street engine using a head gasket with a compressed thickness of 0.035in, a piston-to-deck clearance of 0 would be about right (0.005in more would be better). Keep in mind that a compression ratio loss will not occur at higher engine speeds because of the stretch in the connecting rods.

Honing

When blueprinting an engine, honing serves two purposes. First, as in a standard overhaul, honing is done to remove the glaze from the cylinder walls. As the engine is run over the years, the old piston rings polish the walls of the cylinders.

Engine blocks come in all sizes and shapes. All have the same needs and requirements during an overhaul. A unique aspect of most large diesel engines, like this 60 Series Detroit, and some small gasoline engines are wet sleeves. Wet sleeve engines have removable cylinders.

Honing breaks up this glaze and gives the rings an abrasive surface to help seat them. When blueprinting an engine, honing also ensures that the cylinder walls are as round and straight as possible.

A correct honing technique will produce a cross-hatch pattern on the cylinder walls. Only a flat stone hone can be used to true the cylinder walls. Although a flat stone hone mounted in a drill motor will do a reasonably good job for a stock engine, maximum precision for performance can only be achieved with your machinist's honing machine.

There is a "home" version of a precision honing machine that does an excellent job. If you have a lot of time and a lot of money, or if you are going to be building a lot of engines, then investing in a "home" honing machine might be worthwhile. Proper performance honing requires that the block be submerged in oil during the process. Along with the honing machine, buy an old bathtub and several gallons of honing oil. Also, since there is likely to be 0.004 to 0.006in of metal to remove from each cylinder, be sure to verify that there is nothing you want to see on television for the next week.

If you are having the block bored, it is standard procedure for the machinist to then hone the cylinders. The boring machine will bore the cylinders 0.004 to 0.006in undersize. Honing is then used to remove the remaining metal for proper bore size. It would not hurt to confirm that honing will be part of the procedure when you drop the block off with the machinist.

Rod and crank clearance to the block

When something other than the stock crankshaft or connecting rods are used, there may be problems with them hitting the inside of the block as the engine runs. The crankshaft and rods should have a minimum clearance of 0.050in to all parts of the block. If the clearance between the reciprocating and rotating components is insufficient, contact can occur between these parts and the block. A die grinder should be used to ensure proper clearance of these parts.

Installing the pistons and the camshaft in the block

Installing the connecting rods to the pistons

With the main bearing caps torqued in place, it is now time to install the rods into the pistons. It is critically important to make sure the piston and rod are mated properly. Each has a front and a back to it. Normally there will be a notch in the top of the piston indicating how the piston is oriented to the front of the engine (notch to the front). The rods should have been marked during disassembly. If new rods are being used, refer to the instruction sheet in the box. If the rods were purchased from a machine shop, ask them.

There are two ways the wrist pin is installed into the piston and rod: full-floating and semifloating.

Full floating

Installing the pistons onto the rods in the full-floating wrist pin design is easy, since the wrist pin slides through both the piston and rods with pressure applied by even the frailest of thumbs.

Semifloating

The semifloating wrist pin is more of a problem. The wrist pin either slides through the pistons and presses into the rod, or the wrist pin presses into the piston and slides through the rod. Either one of these offers a perfect opportunity for the rebuilder to damage the piston. The safest route is to have the machine shop install semifloating wrist pins.

A press tool can be made to facilitate the pressing of wrist pins in semifloating applications. Using a piece of hardwood, carve a cradle the

Attention to seemingly innocuous detail is essential when inspecting the block during an overhaul. These fragments of crud (technical term) could restrict coolant flow through the block and cylinder head. This could cause localized overheating in the engine.

diameter of the piston. This will support the piston evenly as the wrist pin is pressed into place. These cradles can also be purchased commercially. Of course, allowing the machinist to install the wrist pins on semifloating applications may be the best answer.

Ring end gap

Place each of the piston rings in the cylinder in which it is to be used. With a feeler gauge, measure the gap between the ends of the rings and compare it to the specifications for this application. If the ring end gap is incorrect, you have the wrong rings or the cylinders have been bored or honed incorrectly.

Installing the rings

If you have not done so already, be sure that the ring grooves are clean and clear of debris and carbon, using a special tool called a ring groove cleaner. However, for those who have good health insurance and a first aid kit, a broken ring can be used to scrape the carbon from the grooves. Extreme caution should be used when using a ring to clean the grooves. The broken ring is sharp and can leave nasty cuts. Wear gloves.

With the ring grooves clean, install the new rings on the pistons. Stagger the rings before installing the pistons into the cylinder. Staggering will provide for a better seal during the early stages of running the engine. The rings are usually identified top, middle, and oil control by the papers covering them in the box. There will also be a mark on the two compression rings indicating which way is up. This mark may be a dot, a bar, or even a corporate logo. In a few instances the rings may be unmarked; there will be instructions in the box.

Two techniques are available for the actual installation of the rings. Start one end of the ring in the correct groove. Rotate the ring around the center of the piston and it will practically fall into place. Plan on breaking an occasional ring using this technique. Piston ring installation pliers can be purchased for only a few dollars at most auto parts stores. Using these pliers virtually eliminates the possibility of breaking rings.

Installing the pistons

With the rods installed in the pistons and the rings in place, put a piston ring compressor around the rings. There may be no piston ring compressor that really works well, so good luck.

Large "industrial strength" cleaning machines like these do an extremely good job of cleaning engine blocks and other components.

With the ring compressor in place, protect the crank journal and the rod bolts from one another with a couple of pieces of rubber hose. Long pieces of rubber hose can even help guide the connecting rod into its proper position on the crankshaft. Gently lower the piston into position. With a small, soft mallet, tap the edges of the piston ring compressor to ensure that it is firmly seated against the deck of the block. Tap the piston firmly into position with the blunt end of the mallet. If the piston is stopped by a ring slipping from between the ring compressor and the block, start again. After installing the first piston, plastigauge the journal. Use the same method described earlier to

Use a small wire brush to clean the openings of the water jacket passageways.

This is the cutting tool used to bore the cylinders. This tool cuts the cylinder to the correct trueness and to slightly less than the correct bore. The machinist will then hone the cylinder to the correct exact bore.

Bearing caps are usually marked to show the correct direction of installation. If there is no marking on the cap, then the locking tabs of the bearing should be matched together.

plastigauge the main bearings. If the oil clearance on the first rod journal is correct, proceed to the next piston. Repeat this process until all the pistons are installed. It may be a little obsessive-compulsive, but it would be a good idea to plastigauge each rod journal before proceeding to the next piston.

After the pistons are installed and the connecting rods are torqued, check the connecting rod side clearance for each rod. Hold the connecting rod as far to one side as possible. Measure the side clearance using a feeler gauge. Verify the acceptable range of clearances with the specifications. Typical side clearances are 0.010 to 0.020in. If the side clearance is outside the specified range for this application, replace the connecting rod. If the side clearance is still incorrect, there is a problem with the crankshaft.

Installing the oil pump

If the oil pump mounts in the oil pan, it should be installed at this point. Many applications mount the oil pump outside the engine, for instance as part of the timing cover; these should be installed later. Remove the oil pump cover and pack a little light grease into it. This will help to prime the pump for initial startup. Torque the cover plate back onto the proper spec. Many OHV applications use the gears of the distributor, by means of a shaft, to drive the oil pump. If your engine is configured in this manner, now is the time to install this shaft. Be sure to inspect the shaft for twisting and straightness; this could be a foreshadowing of future problems. Install the oil pump.

Installing the camshaft

Installing the camshaft on an overhead cam engine is done when the head is assembled. On an overhead valve engine the camshaft is installed into the block and should be done at this time. Installing an OHV camshaft is a lot like pushing a rope uphill. The part that you can hold onto is too short to be able to support it while you thread it through the journals. The answer to this is a camshaft installation tool. These can be purchased or made from a piece of all-thread that matches the center bolt hole of the camshaft. Lube the cam lobes and bearings with assembly lube. Be sure that the assembly lube is a high pressure lube. Forces at the cam lobes can exceed 100,000psi. Using the all-thread to shift the center of gravity and as a handle, install the crankshaft. Care should be taken to ensure the cam lobes do not damage the cam bearings.

Installing the timing gears

On the overhead cam engine this procedure is not usually done until the cylinder head is installed. On the OHV engine the timing gears should be installed now. Typically there will be marks, usually dots, on both the cam gear and the crank gear. These dots normally will line up opposite one another and as close together as possible. There is some variation in this industry standard, and the factory service manual for that engine should be consulted when in doubt.

If the engine uses a chain to drive the camshaft, the gear can be placed in the chain, the crank gear started first, then the crank gear, cam gear, and chain all slide into place together. Usually only gentle persuasion from a soft hammer is necessary to position the gears. In many cases the larger gear, the cam gear, has nylon teeth. The nylon teeth can be damaged by a hammer—be careful.

Where meshed timing gears are used with no chain, the cam gear is usually fiber. Part of the factory tool kit with these applications is a fiber gear puller and installer. Removing the gear with the engine was no problem because you did not care about the condition of the gear. The center of the cam gear where it mates with the camshaft is usually metal. Drive the gear into place using a round, hollow driver on the metal area. As the cam gear slides into place, it will be necessary to be sure the cam can rotate freely as the gear slides onto the shaft. Should the gear bind, it can damage the teeth as it slides into place.

Cam timing

Cam timing is critical to peak performance from a performance engine. To time the camshaft it is first necessary to identify the exact point of TDC for the number one cylinder on the compression stroke. A piston stop is required to do this job. Piston stops can be purchased commercially or can be manufactured from a piece of angle iron and a few bolts. Install a degree wheel on the front of the crankshaft. Position the degree wheel so it is near the top dead center mark when the number one piston appears to be near top dead center. The adjustment is not critical at this point. A pointer to mark TDC can be manufactured from a coat hanger or other trash laying around the shop. Set the piston stop so the piston will only travel within 1/4in to 1/2in of the top of its travel. Rotate the crankshaft clockwise until the piston comes against the stop. Note the reading on the degree wheel. Rotate the crankshaft counterclockwise until the piston comes against the stop. Note the reading on the degree wheel. Top dead center is halfway in between. Bring the crankshaft back to TDC and readjust the degree wheel to read zero.

Install a dial indicator on the intake lifter and rotate the crankshaft against normal rotation until the dial indicator is at its most extended reading. Now rotate the crankshaft with normal crankshaft rotation. When the valve lift reaches the point recommended by the camshaft manufacturer, the degrees shown on the degree wheel should also match the readings recommended by the camshaft manufacturer.

Cast iron or aluminum

There are many substances that can be used as material for the construction of the engine block. For decades the traditional material has been cast iron. Now we could come up with all kinds of reasons why this has been the traditional material. We could say that cast iron is all they had when the internal-combustion, spark-ignition engine was developed. There is a certain amount of truth to that. However, that statement denies the

continued to page 144

These timing gears appear to be installed properly. Notice how the dots on the cam and crank gear line up with one another. However, is the camshaft properly phased to the crankshaft according to the cam manufacturer's specs?

Locate the exact top dead center using a piston stop tool. This one was made from a piece of angle iron, a couple of old head bolts, and a couple of sockets. The dial indicator is only used to ensure accuracy. Rotate the crankshaft in one direction until the number one piston rests firmly against the piston stop. Note the position of the degree wheel. Now rotate the crankshaft in the other direction and note the position on the degree wheel. Top dead center is located exactly halfway between the two noted positions.

Rotate the crankshaft until the camshaft pushes the intake lifter for the number one cylinder to its highest point. The point of maximum lift should not be monitored visually; for accuracy, it should be checked with a dial indicator.

Above and right
Using the recommendations of the camshaft grinder, loosen the bolts in the previously slotted cam sprocket holes and align the cam sprocket so that the intake centerline matches the recommended position on the crankshaft.

While you are working on your overhaul, be sure to keep the engine well protected from foreign objects. A funny but rather sad story about this happened when I was in trade school. You know how every group of guys has at least one jerk. My class's jerk was a fellow whose father had more money than he had sense. Someone dropped some valve keepers in his near-completed engine. When the engine was started, it immediately self-destructed. All that work was wasted. Could not have happened to a nicer guy.

continued from page 141

fact that a great deal of research has gone into improving the metallurgy of the cylinder block without looking at alternative materials.

Cast iron has a relatively small thermal expansion rate. Aluminum, magnesium, and other alternatives have rather large expansion rates. In the United States the operational temperature of the automobile can range from -60 degrees F to over 260 degrees F. This is a temperature differential of 320 degrees F, or about 178 degrees Kelvin. The coefficient of linear expansion for cast iron is 10.5 X 10^{-6}/degree K. The coefficient of thermal expansion is three times the coefficient of linear expansion. A 1in-long bar of cast iron at -60 degrees F will expand by 0.005in by the time the temperature reaches 260 degrees F. Aluminum is another fairly common material for the construction of engine blocks. The coefficient of linear expansion for aluminum is about 23 X 10^{-6}/degree K. That means that if the 1in bar were made of aluminum, it would expand by 0.012in. More than twice as much expansion.

Okay, I can hear you thinking it . . . I live in Atlanta, I never overheat my engine, what difference does this make to me? First, the engine manufacturers never know where the engine is going to be used. It may be in Fairbanks, Alaska. A common overnight temperature might be 40 degrees below zero. The operating temperature of that engine is still 212 degrees or more. The temperature changes by 252 degrees in only a few minutes. Yeah, you are still reminding me that you live in Atlanta. A cast iron block can withstand a great deal of overheating; whereas, an aluminum block will expand to the point where component-to-block clearances drop to zero, or less, and catastrophic damage can occur.

So what are the justifications for changing from cast iron to aluminum? The primary justification is, of course, weight. The density of aluminum is a little over one-third the density of cast iron, a considerable savings in weight. Some of this weight advantage is lost in the need to use thicker castings of aluminum than are necessary with cast iron. Additionally, the melting point of aluminum alloys is less than 1,220 degrees F. Compare this to almost 2,200 degrees F for cast iron. Since combustion temperatures often exceed 2,000 degrees F, any spot where the aluminum block might come in contact with these temperatures must be coated with a heat-resistant material. Again, a loss in the weight advantage.

Typical aluminum block design

There are two typical aluminum block designs. The first is the pressed sleeve, the second is the wet sleeve. A pressed sleeve engine features an aluminum casting similar in appearance to a cast iron block. A steel liner sleeve is then pressed into the bore of each cylinder. Whereas this liner is typically replaceable, damage severe enough to warrant the replacement of the liner is damage severe enough to warrant the replacement of the entire block.

The wet sleeve block looks a lot like the outer part of a cast iron block with the inner parts missing. Individual cylinder sleeves are then set into the block. These sleeves are typically held in place by an interference fit between the block, the liner, the cylinder head, and a brass or paper shim placed under the liner. Many European engines, including popular Formula racing engines, use this technology.

These engines have been rebuilt and are being stored for final delivery.

A finished engine. Satisfaction in accomplishment.

Chapter 8

Lubrication and Cooling

The lubrication system

The lubrication system of the modern gasoline engine consists of a semipressurized oiling system. Most of the moving parts in the engine receive lubrication under pressure from the oil pump; but for many, the lubricating is handled by little more than gravity.

I have heard it argued that oil does not deteriorate or breakdown, that the only deterioration is in the additives. I suppose that if the filtration system were more ideal, that if the environment the engine were being operated in were ideal, and that if there were no leakage of combustion gasses past the pistons and rings, this would be true. Poor maintenance is the most common cause of engine damage. This damage does not occur suddenly in most cases. Slowly, oil volume passing through the engine is reduced. This reduction occurs as the oil pump gears wear and as the screen on the oil pickup becomes restricted with sludge. Long before the reduced oil volume causes a pressure problem, the reduced volume itself causes a problem. The main bearing, rod bearings, and cam bearings are usually the first components to oil. These components will probably receive an adequate, though not ideal, amount of oil. Camshaft lobes, lifters, pushrods, and rocker arms are oiled last. The pistons,

The engine oil removed from an engine can tell the rebuilder a lot about what is wrong with that engine. A quick glance can reveal gray particles of bearings and bright shiny particles of hard metal components. A detailed oil analysis from a laboratory can tell the rebuilder what part of the engine is defective and even how deep into the multilayer bearings the wear has occurred.

cylinder walls, and valve guides are oiled only with the residual oil splashing or dripping from other parts. Since a little of the oil is lost during each oiling stage, reduced oil volume can cause reduced oiling and excessive wear in the last components to be oiled.

It is no accident that parts like the cam lobes, pushrods, and rocker arms were chosen to be oiled after the main bearings and rod bearings. As soon as the main bearings and rod bearings lose oil, they begin to make noise. I once had an apprentice who rebuilt a Volkswagen Rabbit engine. Unlike many engines, only one of the rod bearing shells had a hole to allow oil from the inside of the

This oil was actually removed from an engine belonging to the Alaska Department of Transportation.

This oil pan reveals years of neglect. Delayed oil changes, perhaps low-quality oil, have all worked together to cause premature failure of the engine.

Often poor oiling after a recent overhaul can be caused by insufficient attention being paid to the connections and gaskets between the oil pickup and the oil pump itself.

crankshaft to feed to the bearing-journal surface. My apprentice installed the bearing shell incorrectly. Immediately upon starting the engine a severe rod knock was heard. We shut the engine off, pulled the oil pan, and found no damage to the bearing surfaces. We put the bearing shells in their correct position and there was no problem. The point is, when there is a catastrophic and sudden loss of oil pressure to the main and rod bearings, there is a lot of noise all of a sudden. Sometimes there is little or no damage. A gradual loss of volume is more insidious. Slow oil starvation causes silent wear and silent destruction.

Frequent oil changes reduce the likelihood of sludging; this increases the likelihood of long engine life.

Oil flow in a modern internal-combustion gasoline engine

Oil pan

The oil pan acts as a reservoir, and in most engines the only cooler the oil will have. Although for most domestic passenger car engines the oil pan will hold 4 to 5 quarts, I have worked on applications with a capacity of 12 quarts. The oil pan is shaped to ensure that as lateral and longitudinal "G" forces change, the oil will stay near the oil pump pickup.

Windage trays

Windage trays are virtually required on applications intended for outperforming the car in the next lane. The windage tray separates the engine oil from the rotating crankshaft. Without a windage tray, the crankshaft must rotate through the mist of oil that pervades the air immediately above the oil as it lays in the pan. This mist is created in part by the air currents and eddies created by the rotating crankshaft itself. The addition of the windage tray keeps the oil mist to a minimum by isolating the oil from the rotating crankshaft. In addition, the oil mist, what little is formed, is kept mostly below the windage tray. The less the crankshaft counterweights have to contend with this mist, the less power is lost to the aerodynamic resistance of this action.

Inadequate maintenance and poor-quality oil can cause the oil pickup strainer to become plugged.

Spare no expense on oil pumps. I have seen many amateur engine builders spend hundreds and thousands of dollars on pistons and crankshafts, then install a stock, low-volume, low-pressure oil pump like this one.

The oil

The purpose of the oil is to provide a cushion or film of oil molecule beads between pieces of metal that are in relative movement. Without this layer separating the metal surfaces, extreme friction would result if these metal surfaces came in direct contact.

Oil types

Oil types are described in several different ways. The decision of what type of oil to use should be based on several factors, including those listed below:

Type of engine
Normally aspirated spark ignition
Normally aspirated diesel
Turbocharged spark ignition
Turbocharged diesel
Usage
Usual loads
Operating temperatures (sump temperature)
Amount of combustion residue (blow-by)
Engine details
Oil capacity of the engine
Intended oil change intervals
Type of oil cooling

SAE viscosity ratings

SAE viscosity ratings are intended only to indicate the viscosity of the oil. SAE ratings do not indicate the quality of the oil. The higher the SAE rating, the more viscous or thicker the oil. Generally speaking, if it is winter, or in cold climates, you will want to use a low-viscosity oil. In the summer, or in warm climates, use a high-viscosity oil. As the engine wears, the bearing clearances get greater and the oil pressure drops. A temporary bandage for this can be provided

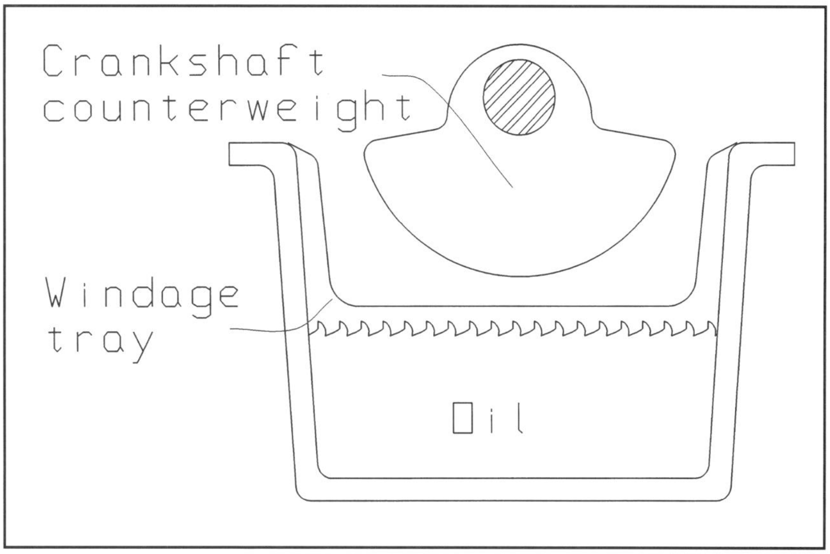

Windage trays are virtually required on applications intended for outperforming the car in the next lane. The windage tray separates the engine oil from the rotating crankshaft. Without a windage tray, the crankshaft must rotate through the mist of oil that pervades the air immediately above the oil as it lies in the pan.

by using an oil with a higher viscosity rating.

SAE 5 Wt.—This is an extremely thin oil recommended only for cold weather and very light load applications. The low viscosity keeps the oil thin at cold temperatures. At extreme sub-zero temperatures, 5 Wt. can allow for free movement of the internal engine parts; that can make the engine easier to start. Since there is a tendency for viscosity to decrease as it heats up, an oil this thin can be too thin to provide proper lubrication when the engine warms up. (A personal note about this: There is a legend that the mechanic who lives on your block will always have the worst running car. It may also be true that the automotive writer on your block may be the one to adhere least to his own advice. As I have previously mentioned, I am living in Fairbanks, Alaska, as I write this book. Several mornings during the past two winters the temperature has plunged to almost 50 degrees below zero. On these mornings especially, I watch the oil pressure gauge on my 1977 Suburban after I start the engine and wish that I had installed winter oil. For several minutes after start-up the oil pressure gauge remains on zero. And, as I write this passage in this book, I am promising myself I will change the oil before the next deep cold, but I know I will not . . . after all, it will be spring again in a few months.)

SAE 10 Wt.—Also extremely thin, SAE 10 Wt. provides slightly better protection than SAE 5 Wt. for a warm engine.

SAE 20 Wt.—For most applications, this is low-viscosity and practical, especially in cold climates, for the passenger cars or light-duty trucks.

SAE 30 Wt.—A good all-around choice for most North American climates. Thin enough to provide easy cold weather starting, assuming of course that you do not spend your winters in Minot, North Dakota, or Fairbanks, Alaska, yet thick enough to provide good protection.

SAE 40 Wt.—An oil with this high a viscosity tends to cause difficult starting when the engine is cold. The thick oil impedes the rotation of the engine during a cold start. Of course, it might not be impractical if you spend your summers in Yuma, Arizona.

Multigrade oils—Multigrade oils have been used for decades to satisfy the needs of both cold weather and hot weather operation. These oils cover three or more viscosity grades. The multiviscosity nature of these oils is accomplished by viscosity index enhancers. The viscosity index relates to an oil's ability to alter its viscosity according to the engine's needs as its temperature increases. A 10W-40 oil would have approximately the same viscosity as 10 weight oil at 0 degrees F. The "W" following the ten indicates that the viscosity test was done at 0 degrees F. The "40" indicates that the viscosity of the oil was tested at 212 degrees F and its viscosity was the same as 40 weight oil.

Viscosity index enhancers often consist of oil-soluble polymers. These enhancers tend to be unstable. The oil should be replaced at the manufacturers' suggested intervals, as these enhancers tend to break down, allowing the oil to thin out at higher temperatures.

American Petroleum Institute grading—American Petroleum Institute grading goes from SA to SG. The API grading system is based on the types of additives in the oil. SA oil is pure mineral oil and should only be used in the lightest duty applications. The higher the second alpha digit, the greater the ability of the oil to protect engines.

Most oil pans are made of stamped steel. Some applications are made from cast aluminum and even plastic. On some engines the oil pan will have an odd shape, a bubble or indentation in the side. This apparent deformity has been engineered into the oil pan to reduce engine noise.

Synthetic oils—These oils have received exaggerated good press and exaggerated bad press. The fact is that synthetic oils have excellent low-temperature characteristics and a better viscosity index than mineral oils. This does not make them a miracle solution to engine wear problems, just good. Their major disadvantage is their expense. A good quality mineral oil with additives appropriate to the engine is more than adequate and considerably cheaper.

Over the years, I have been asked what is the best brand of oil. My usual reply is, "the brand your father used." All name brand oils meet standards appropriate for their use.

Oil pump—Generally the oil pump is a gear pump located in the oil pan. Suspended from the bottom of the block, the oil pump's pickup extends to near the bottom of the oil pan. Here the oil pump picks up engine oil. Over the end of the pickup there is a screen. This screen prevents large particles, nuts, bolts, washers, and other chunks from making their way to the oil pump. This screen can become coated with carbon, residual deposits that result from infrequent oil changes. The residue can limit the ability of the oil pump to pick up the oil at its source.

After being picked up from the oil pan, the oil is carried through the gear teeth of the oil pump. The oil pump's responsibility is to supply the engine with an adequate volume of oil. Oil pressure is determined by the restrictions created by the small clearances between the lubricated surfaces. Should the oil clearances get larger, as a result of wear or damage, the amount of restriction decreases and the oil pressure drops.

Low oil pressure can also result from a lack of volume. Low oil volume can be caused by restrictions in the pickup screen, wear in the oil pump, or restrictions in the oil filter or oil galleries. The low oil pressure that results from the low volume causes excess wear in the bearings and other metal-to-metal surfaces. This wear results in lower pressure. What works best to avoid this scenario is regular maintenance.

The oil pump is also equipped with a pressure relief valve. On new engines or engines with little wear, the potential oil pressure is much higher than the components such as the oil filter can handle. Too much oil pressure can literally blow the oil filter off the engine. The pressure-relief valve limits the oil pressure by diverting some of the oil back to the inlet side of the pump.

A typical oiling path

Once the oil leaves the oil pump, it passes through the oil filter on its way to the main oil gallery. From the main oil gallery the oil is then distributed to the camshaft and lifter. On a pushrod engine the oil is carried through the pushrods to the rocker arms and the upper valvetrain. A second oil path parallel to the upper valvetrain carries the oil to the main bearings of the crankshaft. Oil is then forced through galleries in the crankshaft to the rod bearings. The timing chain on most North American engines is splash oiled. Some applications spray a pressurized stream of oil on the timing chain and gears. Overhead cam engines that use a timing belt do not oil the belt.

Digression

Several years ago I did a technician upgrading class at a large Marine base. They related a story to me about a training session concerning maintenance on tactical vehicles. Apparently this training session had been done in a desert environment. During the training the vehicles had been test driven without the air filters installed. The result was the need to rebuild the engines. The short test drive had allowed enough sand to enter the cylinders to do severe damage to the engine.

This little anecdote illustrates that the air filter is an important

Small details of the cooling system, such as freeze plugs, are often overlooked. This freeze plug looks fine from the outside, but how rusted is it on the inside?

This new set of freeze plugs costs less than $20 and their installation may prevent thousands of dollars worth of damage.

part of the lubrication system. Just as the oil filter removes contaminants from the oil, the air filter helps keep those contaminants from getting in the oil in the first place. As part of the routine maintenance on an engine, the air filter should be replaced. On a throttle body injected and carbureted engine, the little paper gasket between the air cleaner and the top of the carburetor or throttle body is not a throwaway item. This gasket is an essential part of protecting the cylinders from fine particle damage.

Tools like this can be used prior to disassembly to check for cracks in the water jacket of the engine to be rebuilt. They are not a definitive answer and should be used only as a first step in checking for cracked heads and blocks.

This is a pressure-testing machine. It can be used to pressurize the water jackets of heads and blocks. This is an expense that should not be skipped over when you are building a performance engine, or any engine.

When I was in trade school I remember the instructor telling me that two aspirins worth of grit can instantly cause the same amount of wear as 75,000 miles of clean driving.

Dry sumping the engine

It was mentioned earlier that the counterweights of the crankshaft and even the journals of the crankshaft swinging through the oil mist of the crankcase are impeded. This robs power from the engine. In true performance engines, the builder will install an oiling system that will prevent the oil from being stored in a pan below the crankshaft. Oil is kept in a separate and external container. A pump ensures that the oil is transported from the oil galleries and "pan" to the cooler and then to the storage container.

Cooling system

A mistake commonly made, even among professional mechanics, is to spend several hundreds or even thousands of dollars on rebuilding an engine while ignoring the cooling system. This is a serious mistake. Even when the problem that initiated the rebuild is obviously poor lubrication, the cause of the poor lubrication may be poor cooling of the engine block. Every engine rebuild should be accompanied by a thorough inspection of the radiator, hoses, and coolant passages in the engine. Since many of today's radiators feature lateral rather than vertical coolant flow, it is virtually impossible to check the condition of the radiator core visually.

When dozens of hours and hundreds of dollars are being spent to recondition an engine, it is only logical to spend an extra hundred or so to ensure that your work will not be ruined by overheating.

About coolant and its additives

In 1977, I bought an old house in Fort Worth, Texas. In the garage I found a text called *Automobile Engines* with a copyright of 1945. In those days there were three major categories of antifreeze: alcohol, glycerin, and "Eveready Prestone."

Alcohol was the cheapest but had the disadvantage of having a lower boiling temperature than water. Anytime the engine approached what would be considered operating temperature on today's engine, the alcohol would boil away. This made the use of alcohol as a summer coolant impractical, not to mention combustible.

Radiator glycerin had a boiling point higher than water, making it practical in both winter and summer. The primary disadvantages were price and the fact that it did not offer quite as good a protection against winter freeze-up as alcohol.

Eveready Prestone, predecessor to modern antifreeze/coolant

The primary ingredient in modern antifreeze coolant is ethylene glycol. The boiling point of ethylene glycol is 330 degrees F. It is noncorrosive, has no significant odor, and in a 40 to 60 percent mix with water can protect against freeze-up at temperatures as low as minus 65 degrees F. Various additives are added by manufacturers to inhibit leaking, rust, and restrictive scale.

Ethylene glycol is a hazardous waste well known for its ability to do severe kidney and brain damage. Consumption of as little as one-half cup can cause irreversible damage or is fatal. Check with your local department of ecology to determine the proper way to dispose of antifreeze/coolant waste. Look in the future for ethylene glycol to be replaced by a less hazardous chemical. Two of the chemicals being discussed are propylene glycol and polyethylene glycol. While these sound like virtually the same chemical being used today, it is not as toxic. In fact, polyethylene glycol is an ingredient in popular soft drinks.

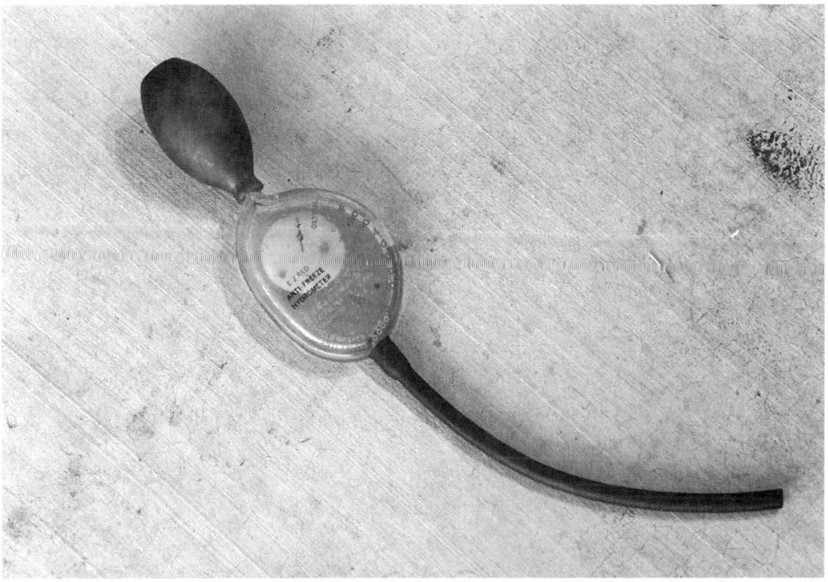

An important part of routine maintenance is to check the "richness" of the coolant. I live in Fairbanks and keep mine at about -45 degrees F. In most places in the country, -20 degrees F should be sufficient.

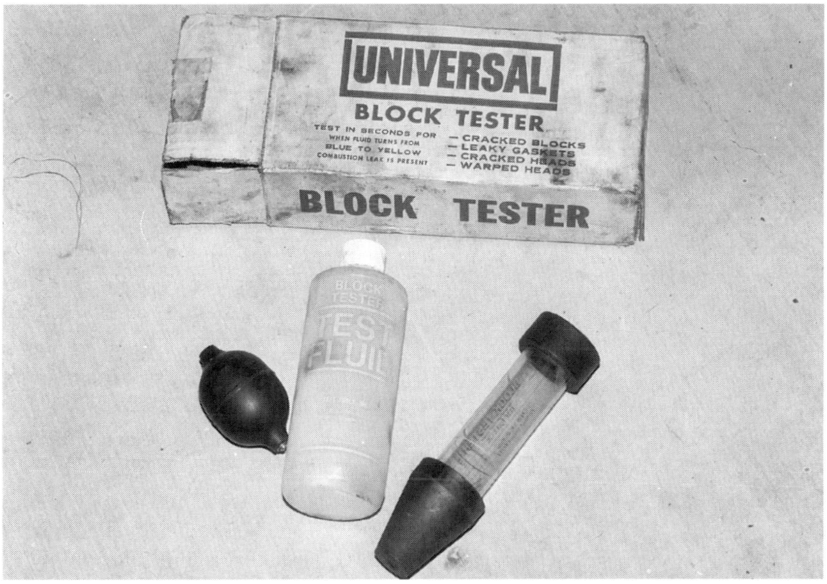

If a cracked cylinder wall or blown head gasket were suspected prior to disassembly, a block checker can be used to confirm the condition. Filled with a blue fluid and inserted into the filler neck of the radiator, the chemical block tester extracts some of the vapor in the radiator. The presence of the hydrocarbons in the radiator vapors indicates leakage from the combustion chamber into the radiator. This leakage could be a result of a blown head gasket, cracked cylinder, or cracked cylinder head. If hydrocarbons are present in the vapors, the blue fluid will turn yellowish-green.

Thermostats are sometimes left out on performance engines. Be aware that this practice can sometimes cause an effect opposite of that which was desired. The coolant may move through the radiator so fast that it does not have time to exchange heat to the air passing through the radiator. The result can be a severely overheating engine. Again, I speak from experience.

Components of the cooling system

Radiator

The radiator is a heat exchanger made of copper or aluminum. Older radiator cores consist of copper tubes separated by corrugated fins. The fins and tubes dissipate the heat of the coolant to the air passing through the radiator. These radiators have copper tanks located at each end of the core. Back when my dad was trying to teach me how to release a clutch without leaving the universal joints in the gutter, the tanks were located on the top and bottom of the radiator. This was called a vertical flow radiator. Most radiators today are of the cross-flow design. In the cross-flow design, the tanks are located on the left and right sides of the radiator.

Many modern radiators consist of aluminum core tanks made of polyamide with glass-fiber reinforcement. One distinct advantage of the polyamide tanks for the manufacturer is that the mounting brackets can be cast as part of the tank. This reduces the vehicle's production cost. Some of the German imports were among the first to use this design. In the early days of these "plastic tank" radiators, we had a lot of leaking problems. I have a prejudice against them to this day. In spite of this, like many technologies, they have improved over the years. They seem to be quite dependable today.

Transmission oil cooler

Most cars equipped with an automatic transmission have a transmission oil cooler in the radiator. Sometimes oil from the transmission cooler can leak into the radiator. This oil can sometimes be misinterpreted as a blown head gasket or intake manifold problem.

Hoses

Water hoses are the part of the cooling system that the car owner has the greatest ability to monitor. Good quality hoses can be expensive but are worth every penny. Over the years I have used the corrugated, so-called universal radiator hoses. It may be personal prejudice showing through again, but today's engine compartments are so tight and the angles that the hoses have to be twisted into are so sharp, pop for an extra few bucks and buy the molded hoses. To be perfectly frank, when I told my customers, "Of course it's as good as a molded hose, and cheaper," what I really meant was, "I can't get the right hose for your car."

Many automotive fleet users replace the standard black rubber radiator and heater hoses with green silicone hoses. These hoses are most commonly found on emergency equipment applications. Although they are more expensive, they are also much more durable.

Years ago I bought a Pinto that nearly cost me my marriage. One mild North Central Texas afternoon (the temperature was only in the low hundreds), I blew a lower radiator hose. The coolant dumped out of the cooling system so fast that I did not even notice. Fortunately, I had a scapegoat for my inattention, a corrugated universal hose. While these hoses are useful for emergency repairs, they are not nearly as dependable as the preformed hoses.

Hose clamps

There is no such thing as a used hose clamp. A new 50 cent clamp can mean the difference between a successful, long-lasting overhaul and a $1,000 box full of new, but junk, parts.

Radiator cap

You can buy a new one for about five bucks, so it cannot be important, right? Wrong! The radiator cap is both the pressure regulator and the pressure relief valve for the cooling system. If the cap does not hold a high

This trash bin is filled with neglect and shortcuts.

enough pressure, coolant loss and overheating can result. If the cap does not properly relieve pressure, the tanks can literally blow off the radiator.

Water pump

The water pump is responsible for ensuring that the coolant moves through the cylinder head and block to remove potentially damaging heat. Further, it moves the heat-saturated coolant through the hoses to the radiator where the heat can be dissipated to the air passing through the radiator. Although the water pump is normally driven by belts on the front of the engine, the SAAB engine drives the water pump with the camshaft.

Thermostat

The thermostat prevents coolant from passing through the radiator until the engine reaches operating temperature. It is then responsible for ensuring the coolant temperature does not drop below operating temperature.

It is common practice in places like Guam to remove the thermostat entirely. While this may seem like an incredible feat of wisdom for a warm climate, in reality, removing the thermostat may allow the coolant to pass through the radiator so fast that the radiator does not have a chance to cool it. Removing the thermostat can actually cause overheating.

The thermostat is a temperature-controlled valve. As the temperature of the coolant behind the valve rises, the thermostat opens, allowing the coolant to pass into the radiator. If the thermostat fails to open, the engine will overheat. Replace the thermostat anytime major work is done on the engine.

Water jacket

The water jacket is the passageway cast into the cylinder head and block through which the coolant passes.

Chapter 9

What's Legal and What's Not

The basic rule when making internal engine modifications for vehicles that are in smog check areas or areas that are likely to be smog check areas in the future is . . . don't! The law in California, which we are forced to use as the strictest standards, permits no internal modifications for domestic cars after 1966, and for foreign cars, no modifications after 1968. There is the temptation to hide the fact that high-compression pistons have been installed in the engine. After all, when the emissions are tested, they are not going to take the engine apart to determine if the stock pistons are installed. While that may be technically accurate, it is likely that the emissions may be changed enough to fail emissions.

Here in Fairbanks we have a severe problem with CO emissions during the winter. Hydrocarbon (HC) and oxides of nitrogen (NOx) emissions are not a problem. In this jurisdiction it is legal to make any internal engine modifications you desire. But even here in the great white north, there is no immunity to federal intervention. At any point in the future the federal Environmental Protection Agency may come in and require tighter emission standards. The best policy when modifying an engine destined for street use is to have an honest heart-to-heart talk with the I/M referee for your jurisdiction. Most of these people are motorheads anyway. They love to bench race with the best of them. I sympathize in advance with those who end up with a grumpy referee.

In many jurisdictions the depth of the regulation is well beyond what the referee can handle. There is still an avenue for those who have a great deal of time, money, and an entrepreneurial spirit. Let me explain the procedure. Let us say that we have a 1995 Ford light-duty truck that we wish to convert to natural gas. When converted to natural gas, the emissions will be lower (see the next chapter on alternative fuels). These lower emissions come at the expense of power. To boost the power back to about where it was, it would be advisable to increase the compression ratio. Logic and science dictate that the use of high-octane fuel like natural gas in a high-compression engine should not create high levels of emissions. Carbon monoxide, hydrocarbons, and oxides of nitrogen should all remain within their legal limits. The California Air Resources Board is open to that concept. All you have to do is prove it. Modify about four vehicles of the exact type that you wish to modify. Have each of the four vehicles CARB (California Air Research Board) certified. This procedure is much cheaper than it used to be. A typical test will only cost you about $1,500 to $2,000. So, for about $6,000 to $10,000 (not counting the cost of the vehicles), you can have your engine modification certified.

Unlike most scientific regimens, the law has subtle twists, turns, and traps. This chapter is not intended to be legal advice, it is intended to point out some of the issues involved in the high-performance modification of late model cars. The bulk of the information that follows came from California's Bureau of Auto Repair (BAR) and the CARB. The laws on the books in many states are similar to those of California. California has the longest record of stringent enforcement, which is why I am using their information.

There are three categories of replacement parts recognized by the ARB.

Category 1

Category 1 items are not considered by the BAR or ARB to be of any concern as long as the required emission controls are not tampered with. Here is a list of category 1 items.

- PCV air bleeds
- Air cleaner modification
- Air conditioner cut-out systems
- Antitheft systems
- Blow-by oil separators and filters
- Electronic ignition systems retrofitted to vehicles originally fitted with point/condenser systems, as long as the original advance controls are maintained
- Engine shut-off systems
- Ignition bridges and coil modifications
- Throttle lockout systems
- Intercoolers for OEM turbocharger
- Under-carburetor screens
- Vapor/steam/water injectors

Category 2

Category 2 addresses allowable replacement parts.

- Headers on noncatalyst cars
- Heat stoves for allowed headers
- Intake manifolds for non-EGR vehicles must allow for the installation and proper functioning of the OEM

emission controls.

- Approved aftermarket catalytic converters
- Carburetors marketed as "emission replacement"
- Replacement fuel fill-pipe restrictors
- Replacement gas caps

You can see from this list that for catalyst-equipped fuel-injected cars, there are no performance replacements that are allowable without type approval from the ARB. Today's cars are EPA inspected as an integrated system; disturbing even the minutest portion of the emission control package would constitute a violation. Category 3 parts must have verification of acceptability. If you are replacing a part in Category 3, ask for and retain a copy of the verification of acceptability for that product. It may prove handy later on, even if you live in an area that is not currently strictly controlled.

Category 3

Category 3 includes:

- Carburetor conversions
- Carburetors that replace OEM fuel injection
- EGR system modifications
- Replacement PROMs (computer chips)
- Electronic ignition enhancements for computerized vehicles
- Exhaust headers for catalyst vehicles
- Fuel-injection systems that replace OEM carburetors
- Superchargers
- Turbochargers

Chapter 10
Alternative Fuels

We start at a precipice. One of the most exciting events to happen in the world of automotive technology in decades is about to occur. For over a hundred years the fuel of choice for the automobile has been gasoline. Over the next decade there is a good chance that this monopoly will give way to one or more of a plethora of new fuels. Methanol, ethanol, natural gas, and even hydrogen are being touted as viable alternatives to gasoline.

Octane, alternative fuels, and compression ratio

The primary disadvantage of these alternative fuels is their lack of heat energy per unit of mass. Most of them have only a fraction of that contained in gasoline. The good news is that the newer fuels have a much higher octane than gasoline. Higher octane means higher compression ratios are possible without the risk of damaging detonation. In fact, higher compression is not only possible, it is desirable. As the compression ratio is increased, the power lost by the engine is largely regained.

The two most dominant alternative fuels looked at over the past few years are CNG (compressed natural gas) and methanol.

Methanol has for a long time been one of the performance fuels of choice. Its high-octane and relatively high heat content provide excellent performance. However, there are a couple of negatives concerning methanol. First, it is toxic. Second, it is corrosive. Methanol-fueled vehicles must have special and very expensive fuel system components.

Compressed natural gas is the other dominant alternative fuel. In many ways it leads the charge for fuels other than gasoline. Throughout most of the United States there is an infrastructure already in place for distribution of natural gas. It has a low toxicity and a high flash point. Therefore, it is a safe and easy-to-obtain fuel.

Engines built for either of these fuels have needs that differ from those of the gasoline spark-ignition engine. Since natural gas seems to be the leader in alternative fuels at this writing, that is where I will concentrate my discussion. The fuel is slower burning. The timing, therefore, needs to be advanced more at a given engine rpm than the same engine running on gasoline. The timing should also be allowed to stay more advanced at a given engine load condition. These needs are taken care of automatically by many electronic timing control systems. Many of these systems feature detonation sensors that detect pinging caused by high load conditions and low-octane fuel. The computer, upon detecting a detonation signal from the knock sensor, retards the timing. On some of these systems the computer will advance the timing until a detonation is detected. Then and only then is the advance of the timing limited. On applications with this behavior pattern, the natural gas fuel postpones detonation, and the timing advances considerably beyond what detonation will allow it to advance when running on gasoline. The additional timing advance required for CNG is automatically supplied.

Now, to the things that relate to this book directly. Many of the techniques and strategies that have been used in the past to turn a cast iron gasoline boat anchor into a horsepower demon are perfect for turning a gasoline engine into a dedicated CNG engine. Keep in mind, however, that if you choose to make major changes to the engine to run on natural gas, the engine will no longer run well on standard pump gasoline.

Compression ratio

Even with sophisticated detonation sensing and timing control systems, the realistic limit for compression on a street gasoline engine is about 10:1. Higher compression ratios can have such a severe tendency to detonate that power actually tends to decrease with an increase in compression ratio. With natural gas, 10:1 is a good starting point. Power can be increased more by boosting the compression ratio closer to 12:1.

Since CNG is a gaseous fuel, the nature of the fuel itself has a distinct advantage over gasoline. Gasoline tends to "globularize" as it is carried by the air stream toward the combustion chamber. This means there is an inherent tendency for the air/fuel charge to be uneven in the combustion chamber. Gaseous fuels like CNG are carried as an evenly mixed suspension into the combustion chamber. Earlier we discussed the problem of the piston dome interrupting the travel of the flame front though the combustion chamber. This problem is virtually eliminated with CNG. High-domed pistons are quite satisfactory for use in CNG applications, often without providing any relief of the piston dome.

Metallurgy

Natural gas has a higher flash point (ignition temperature) than does gasoline; however, it burns cooler. The lower burn temperature means that the piston can be made lighter and a little less tough. Natural gas engines have been shown to require less maintenance and fewer major repairs than gasoline engines.

Conversions

Natural gas conversions are relatively expensive. As much as $2,000 to $4,000. However, some local gas companies have subsidy plans that can either partially or totally reimburse the car owner for these conversions. Natural gas has some exciting prospects for the future, some exciting prospects for those who have the skills to upgrade gasoline engine designs for this new fuel.

Index